DIGITAL TECHNOLOGIES

FOR CLIMATE ACTION, DISASTER RESILIENCE, AND ENVIRONMENTAL SUSTAINABILITY

OCTOBER 2021

ASIAN DEVELOPMENT BANK

ADB

© 2021 Asian Development Bank
6 ADB Avenue, Mandaluyong City, 1550 Metro Manila, Philippines
Tel +63 2 8632 4444; Fax +63 2 8636 2444
www.adb.org

Some rights reserved. Published in 2021.

ISBN 978-92-9262-879-6 (print); 978-92-9262-880-2 (electronic); 978-92-9262-881-9 (ebook)
Publication Stock No. TCS210188-2
DOI http://dx.doi.org/10.22617/TCS210188-2

The views expressed in this publication are those of the authors and do not necessarily reflect the views and policies of the Asian Development Bank (ADB) or its Board of Governors or the governments they represent.

ADB does not guarantee the accuracy of the data included in this publication and accepts no responsibility for any consequence of their use. The mention of specific companies or products of manufacturers does not imply that they are endorsed or recommended by ADB in preference to others of a similar nature that are not mentioned.

By making any designation of or reference to a particular territory or geographic area, or by using the term "country" in this document, ADB does not intend to make any judgments as to the legal or other status of any territory or area.

Corrigenda to ADB publications may be found at http://www.adb.org/publications/corrigenda.

Notes:
In this publication, "$" refers to United States dollars.

ADB recognizes "China" as the People's Republic of China, "South Korea" as the Republic of Korea, and "Vietnam" as Viet Nam.

Cover design by Edith Creus.

Contents

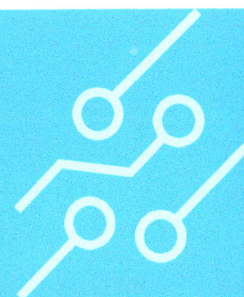

Tables, Figures, and Boxes

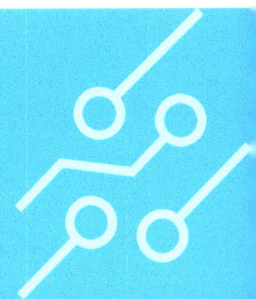

Boxes

Foreword

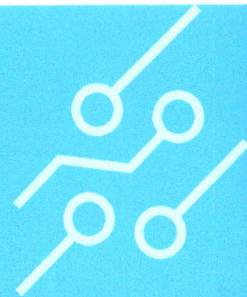

The coronavirus disease (COVID-19) pandemic has created unprecedented health and economic crises. An initial decline in greenhouse-gas emissions due to the pandemic was short lived and the world is not on track to achieve the targets articulated under the Paris Agreement. The arguments for a green, inclusive, and resilient economic recovery are compelling. In this context, well-designed initiatives leveraging the benefits of Digital Technologies (DTs) can be instrumental in helping countries shape low-carbon pathways that are less vulnerable to risks.

There is a wide spectrum of DTs, and they can vary from simple applications, such as mobile text alerts for early and emergency warnings, to extremely complex functions, such as predictive analysis of the probability of natural hazards. The landscape of DTs is rapidly evolving, and recent innovations are yet to be commercialized for wider usage.

Innovative uses of DTs—for climate action, disaster resilience and environmental sustainability—are gaining momentum. In disaster response, for example, the Digital Humanitarian Network in 2012 analyzed over 20,000 feeds from social media to create a geo-located and timestamped database to map the impacts of Typhoon Bopha in the Philippines. This map played a vital role in the initial assessment of the typhoon's impact on key sectors and provinces and planning of immediate relief efforts. Similarly, following the April 2015 earthquake in Nepal, relief workers used drone data and maps provided by volunteer groups to ensure the delivery of emergency supplies to survivors.

The potential of DTs is large and growing. Institutions are embarking upon ambitious DT projects that will transform the way we respond to climate and environment challenges and manage disaster risk. For example, the Destination Earth (DestinE) platform of the European Union (EU) is expected to be implemented gradually over the next 7 to 10 years with the aim of developing a digital model of the Earth that can simulate natural and human activity. DestinE will monitor changes to the atmosphere, ocean, ice, and land cover as well as human activities with a high degree of precision. The EU hopes to use DestinE to accurately forecast floods, droughts, and fires many days or even years in advance, to enable timely planning and action. This platform is also expected to provide policymakers with analyses of the impacts of climate change and the potential effectiveness of climate policies.

Improving the understanding of the potential role of DTs in climate action, disaster resilience, and environmental sustainability, as well as methods for strengthening cybersecurity and protecting data privacy, is timely and important. Countries need support in adopting and implementing policies and regulations that create an enabling environment for harnessing the full potential of new and emerging technologies. This needs to be complemented by appropriate national and international measures to ease access to DTs and promote the transfer of knowledge and technology, particularly for the most vulnerable countries.

This publication presents a broad landscape of DTs at various phases of development and usage that countries can utilize to improve their efforts in tackling climate change, building climate and disaster resilience, and enhancing environmental sustainability. It builds on lessons learned from existing DTs and delves into important considerations such as scale of application, relative merits, cost-effectiveness, and limitations of specific DTs. Finally, it provides insights on how to operationalize DTs considering national priorities and circumstances. We hope that this publication will be useful to countries in selecting DTs that can help achieve their climate goals and foster a green, inclusive, and resilient recovery.

Bambang Susantono
Vice-President for Knowledge Management and Sustainable Development
Asian Development Bank

Preface

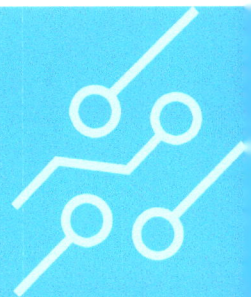

Climate change, environmental degradation, and disasters triggered by natural hazards are among the biggest challenges the world faces today. This is evidenced by increased intensity and frequency of extreme weather events, damage to ecosystems and displacement of livelihoods and people. The record damage caused by cyclone Amphan in Bangladesh and India in May 2020, untenable plastic pollution in the marine ecosystem with eight million tons of plastic entering the ocean every year and mostly from Asia, and over 21 million internally displaced people due to disasters in 2020 in Asia and the Pacific highlight the vulnerability of our economic environment and social systems.

The call to action to address these challenges necessitates pooling of all available solutions including innovative technological, financial, and social responses. In this context, the role of Digital Technologies (DTs) is becoming more and more recognized for tackling climate change, building climate and disaster resilience, and enhancing environmental sustainability, which are the objectives of operational priority 3 (OP3) of ADB's Strategy 2030. This publication presents the broad landscape of DTs and how they can contribute to these strategic priorities.

An exploration of the DTs in various stages is undertaken to understand the opportunities, including those in Stage I which are readily available such as internet and geographic information systems; those in Stage II that are available but are not being utilized to their full potential such as social media, mobile applications, and cloud computing; and technologies in Stage III, such as Internet of Things, Distributed Ledger Technologies, and Artificial Intelligence, which are expected to become commercial in the future and have the potential to significantly accelerate actions to address priorities under ADB's OP3.

Some of these technologies have been usefully deployed in relevant context in countries in Asia and the Pacific region. These include DTs as enablers such as relevant connections between sets of data, communication among environmental sensors and databases, monitoring and tracking changes in ecosystems, analysis of DRM-relevant data sets, optimization of sustainable supply chains, and prediction of vehicular traffic and weather phenomena with high degrees of accuracy. The benefits of DTs accruing through reduced transaction costs and data and information protection is also a highlight as in emissions trading systems.

However, the selection of DTs requires a careful consideration of the larger development objectives weighed against cost effectiveness and limitations, and above all, the local circumstances. Assessments through Marginal Abatement Cost Curves and Marginal Adaptation Cost Curves can facilitate informed consideration of "appropriate technologies" for the developing countries. Due cognizance of the security and privacy issues as well as capacity and usability of individual technologies in the country context is also critical.

This publication also explores conditions that can enhance deployment including creating knowledge and awareness, selecting targeted sectors or user groups, and building public private partnerships. An overarching DT policy framework and investments in hard and soft DT infrastructure are of course the preconditions for any successful deployment. The publication also highlights how investments in green recovery could integrate DTs. We hope that the insights provided in this publication will help countries in making informed choices for mainstreaming rapidly advancing DTs in their efforts to tackle climate change, build disaster resilience, and enhance environmental sustainability.

Xiaohong Yang
Chielf Thematic Officer
Thematic Advisory Service Cluster
concurrently Officer-in-Charge,
Climate Change and Disaster Risk
Management Division

Virender Kumar Duggal
Principal Climate Change Specialist
and Fund Manager, Future Carbon Fund
Sustainable Development and
Climate Change Department
Asian Development Bank

Acknowledgments

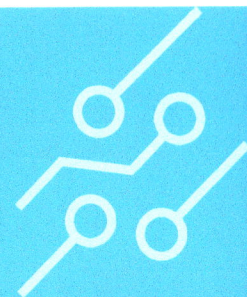

Digital Technologies for Climate Action, Disaster Resilience, and Environmental Sustainability was developed by the Climate Change and Disaster Risk Management Division (CCDRM) within the Sustainable Development and Climate Change Department (SDCC) of the Asian Development Bank (ADB).

Virender Kumar Duggal, principal climate change specialist, Climate Change and Disaster Risk Management Division (SDCD), conceptualized and guided the development of this knowledge product. Preety Bhandari, former Chief of CCDRM and concurrently Director, SDCD encouraged its development and provided strategic direction. Steven Goldfinch, disaster risk management specialist; Arghya Sinha Roy, senior climate change specialist (Climate Change Adaptation); and Takeshi Miyata, climate change specialist, provided insightful inputs for its development.

Georg Caspary of Scaletech Ltd. conducted the research and led the analysis. The development of this publication also benefited from advice and inputs from Brij Mohan, Deborah Cornland, Hannah Ebro, Naresh Badhwar, and Rastraraj Bhandari which is appreciated.

This knowledge product has hugely benefited from the valuable inputs and peer review conducted by Nir Kshetri of the University of North Carolina—Greensboro, which is deeply appreciated. Thomas Abell, Advisor, SDCC and Chief of Digital Technology for Development at the ADB also conducted an insightful peer review, which is sincerely acknowledged and appreciated.

The timely publication of this knowledge product was made possible by the valuable coordination and administrative support of Janet Arlene Amponin, Anna Liza Cinco, Ken Edward Concepcion, Ketchie Molina, Jeanette Morales, Ghia Villareal, and Zarah Zafra.

The preparation of this publication has immensely benefited from the diligent support from Lawrence Casiraya (page proof checking), Edith Creus (cover design and layout), Monina Gamboa (editing), and Jess Alfonso Macasaet (proofreading), all of which is sincerely appreciated.

The support provided by the Publishing Team in ADB's Department of Communications and the Printing Services Unit in ADB's Office of Administrative Services in publishing this knowledge product is also acknowledged.

Abbreviations

4G	fourth generation
5G	fifth generation
ADB	Asian Development Bank
ADFI	African Digital Financial Inclusion Facility
AfDB	African Development Bank
AHP	analytical hierarchy process
AI	artificial intelligence
AR	augmented reality
AV	autonomous vehicle
CCA	climate change adaptation
CCM	climate change mitigation
CO_2	carbon dioxide
DL	deep learning
DLT	distributed ledger technology
DMC	developing member country
DPA	digital process automation
DRM	disaster risk management
DT	digital technology
EC	European Commission
EU	European Union
GHG	greenhouse gas
GIS	geographic information system
GPT	general-purpose technology
IADB	Inter-American Development Bank
ICT	information and communication technology
IDC	International Data Corporation
INDC	intended nationally determined contribution
IPCC	Intergovernmental Panel on Climate Change
IoT	Internet of Things

IS	information system
ITU	International Telecommunication Union
MACC	Marginal Abatement Cost Curve
MAdCC	Marginal Adaptation Cost Curve
MCA	multi-criteria analysis
ML	machine learning
MR	mixed reality
MRV	Monitoring, Reporting, and Verification System
ms	millisecond
NASA	National Aeronautics and Space Administration
NDC	nationally determined contribution
OECD	Organisation for Economic Co-operation and Development
OP	Operational Priority
PPP	public–private partnership
PRC	People's Republic of China
R&D	research and development
RPA	robotic process automation
SC	supply chain
SMART	Spatial Monitoring and Reporting Tool
UAV	unmanned land, sea, and air vehicle
UK	United Kingdom
UN	United Nations
UNDP	United Nations Development Programme
UNDRR	United Nations Office for Disaster Risk Reduction
UNFCCC	United Nations Framework Convention on Climate Change
US	United States
VR	virtual reality
WIPO	World Intellectual Property Organization
XR	cross reality

1 Introduction

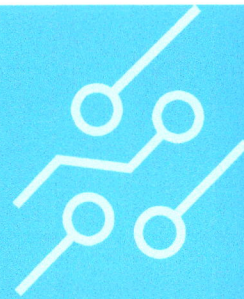

Digital technology (DT) as a development tool provides unique opportunities for developing countries to leapfrog intermediate steps in development while improving the quality and broadening the reach of public services. It can also provide economic and social progress and enhance its effects and help achieve the development targets in the post-2020 world including climate action and sustainable development. In fact, four out of the 17 Sustainable Development Goals make direct reference to DT. The strategic and effective use of DT—combined with a reform-oriented mindset, necessary set of skills, institutional structure and capacity, appropriate business models, as well as policy and regulatory environments—can facilitate fast and efficient delivery of public services in key sectors.[1]

The Asian Development Bank (ADB), through its Strategy 2030, sets out a series of operational plans to set the course for ADB's efforts to responding the changing needs for the Asia and Pacific region until 2030.[2] These seven operational plans contribute to ADB's vision to achieve prosperity, inclusion, resilience, and sustainability, and are closely aligned with Strategy 2030 principles and approaches. Among other priorities, Operational Priority 3 (OP3) focuses on tackling climate change, building climate and disaster resilience, and enhancing environmental sustainability (Figure 1).

ADB's Strategy 2030 also underscores the importance of innovative technology and integrated solutions. Accordingly, ADB is committed to strengthening its country-focused approach using the country partnership strategy as the main platform to define customized support, promote the use of innovative technologies, and deliver integrated solutions by combining expertise across a range of sectors and themes and through a mix of public and private sector operations. This provides a foundation for integrating DT in achieving OP3 as well as Strategy 2030.

DTs have boosted growth, expanded opportunities, and improved service delivery in much of the world. Over the past 2 decades, ADB has extended DT-related loans, grants, and technical assistance projects to help develop and maintain sectors including DT infrastructure, DT industries, and DT-enabled services, DT policy, strategy, and capacity development and it continues to do so. From 2010 to 2019, ADB supported 371 projects that included digital components (including 27 nonsovereign projects) in sectors ranging from telecommunications, education, and health to agriculture and food security. For example, Uzbekistan's Digital Technology in Basic Education Project, supported by a $30 million ADB loan, is bringing education to remote rural areas and benefiting 540,000 students.

DTs can help tackle climate change, build climate and disaster resilience, and enhance environmental sustainability. For example, it can do so by enabling relevant connections between sets of data, enabling communication among sensors and databases, monitoring and tracking changes in ecosystems, analyzing data

---

[1] ADB's Focus on Digital Technology. https://www.adb.org/what-we-do/sectors/dt/overview.
[2] ADB. 2018. Strategy 2030: Operational Plans. https://www.adb.org/about/strategy-2030-operational-priorities. Manila.

Figure 1: Operational Priority 3: Tackling Climate Change, Building Climate and Disaster Resilience, and Enhancing Environmental Sustainability

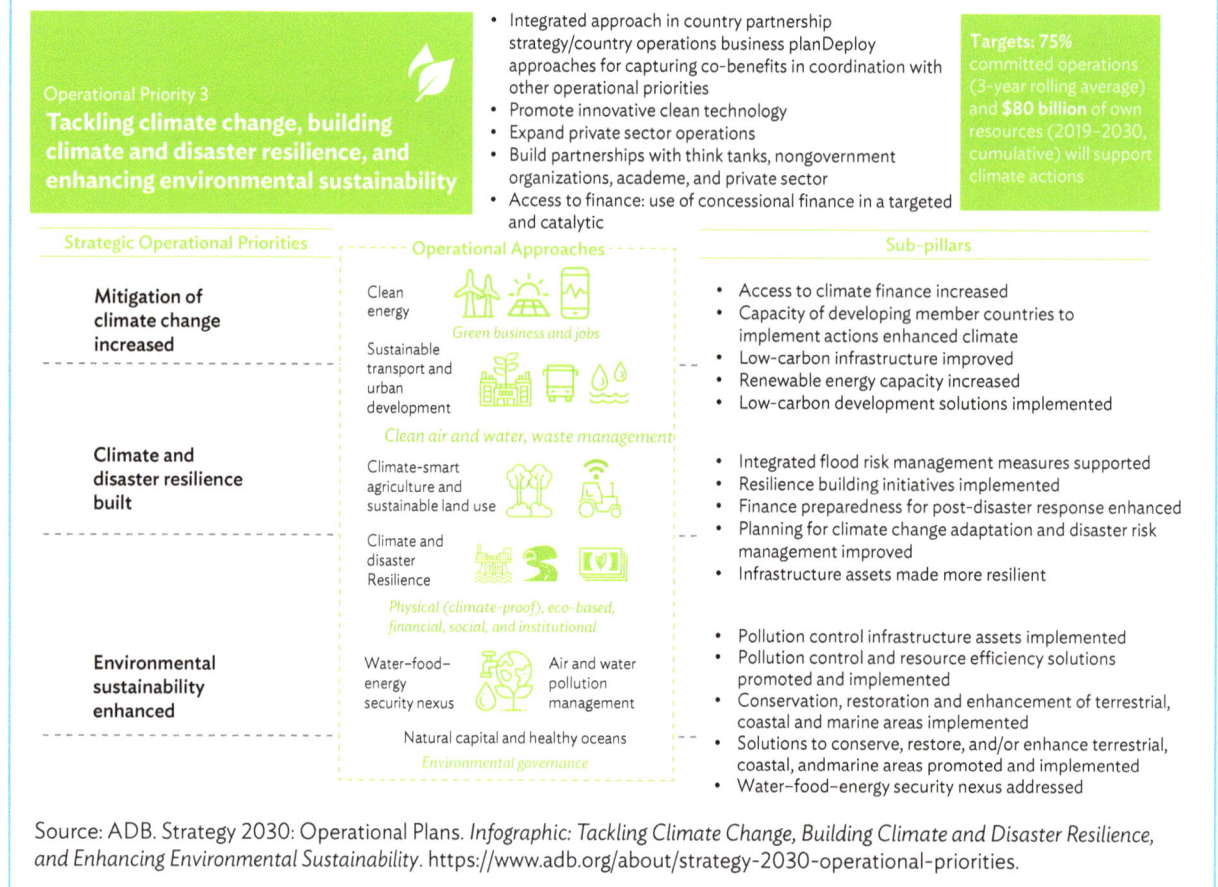

Operational Priority 3
Tackling climate change, building climate and disaster resilience, and enhancing environmental sustainability

- Integrated approach in country partnership strategy/country operations business planDeploy approaches for capturing co-benefits in coordination with other operational priorities
- Promote innovative clean technology
- Expand private sector operations
- Build partnerships with think tanks, nongovernment organizations, academe, and private sector
- Access to finance: use of concessional finance in a targeted and catalytic

Targets: 75% committed operations (3-year rolling average) and **$80 billion** of own resources (2019–2030, cumulative) will support climate actions

Strategic Operational Priorities	Operational Approaches	Sub-pillars
Mitigation of climate change increased	Clean energy / *Green business and jobs* / Sustainable transport and urban development / *Clean air and water, waste management*	• Access to climate finance increased • Capacity of developing member countries to implement actions enhanced climate • Low-carbon infrastructure improved • Renewable energy capacity increased • Low-carbon development solutions implemented
Climate and disaster resilience built	Climate-smart agriculture and sustainable land use / Climate and disaster Resilience / *Physical (climate-proof), eco-based, financial, social, and institutional*	• Integrated flood risk management measures supported • Resilience building initiatives implemented • Finance preparedness for post-disaster response enhanced • Planning for climate change adaptation and disaster risk management improved • Infrastructure assets made more resilient
Environmental sustainability enhanced	Water–food–energy security nexus / Air and water pollution management / Natural capital and healthy oceans / *Environmental governance*	• Pollution control infrastructure assets implemented • Pollution control and resource efficiency solutions promoted and implemented • Conservation, restoration and enhancement of terrestrial, coastal and marine areas implemented • Solutions to conserve, restore, and/or enhance terrestrial, coastal, andmarine areas promoted and implemented • Water–food–energy security nexus addressed

Source: ADB. Strategy 2030: Operational Plans. *Infographic: Tackling Climate Change, Building Climate and Disaster Resilience, and Enhancing Environmental Sustainability.* https://www.adb.org/about/strategy-2030-operational-priorities.

sets, optimizing supply chains, and predicting numbers of vehicles in traffic and weather phenomena. DTs also reduce transaction costs and can protect information from being corrupted or modified, hence, reducing the cost of validation in carbon trading. This publication highlights the landscape of DTs available to countries in addressing challenges related to OP3. Understanding these variety of DTs alongside their scope, accessibility, and relevance, however, remains a challenge given the dramatic growth in DTs over the past years.

It is in this context that this publication presents the landscape of DTs for tacking climate change, building climate and disaster resilience, and enhancing environmental sustainability. It starts by laying out the range of existing DTs and their respective applications in climate change, environmental sustainability, and the disaster risk management arena (Chapter 2). Next, it provides an extensive overview of the role DTs can play in tackling climate change, building climate and disaster resilience, and enhancing environmental sustainability (Chapter 3). After dwelling on several considerations in selecting DTs to address development objectives on climate, disaster resilience, and environmental sustainability (Chapter 4), the publication provides insights on operationalizing DTs in Developing Asia (Chapter 5), and Chapter 6 concludes the publication. An Appendix provides further information on ADB's ongoing work on the application of DTs in its policies for key sectors as well as decision-making.

2 The Landscape of Digital Technologies

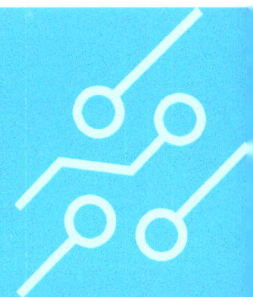

This chapter divides the broad range of DTs (see Box 1 for definitions and explanations) currently available and structures them into three distinct stages. It then discusses their respective applications for each stage in the climate change mitigation (CCM), climate change adaptation (CCA), disaster risk management (DRM), and environmental sustainability arenas. Figure 2 illustrates these three stages and the main technologies in each stage that will be discussed in this publication.

Figure 2: Ecosystem of Digital Technologies

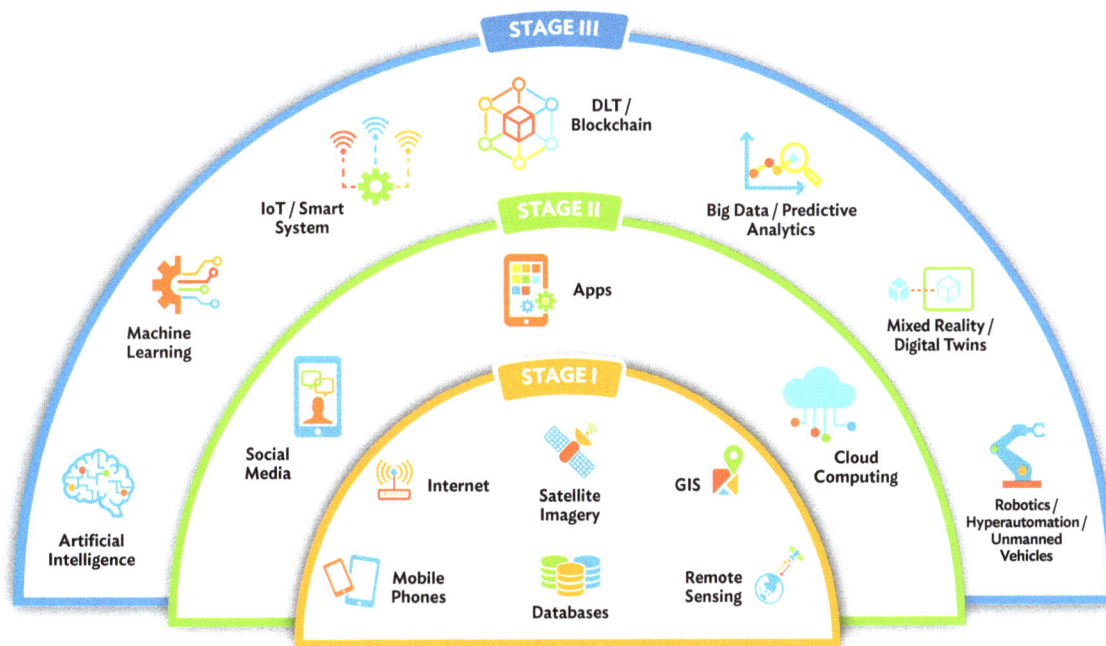

DLT = distributed ledger technology, GIS = geographic information system, IoT = Internet of Things.
Source: Asian Development Bank.

Box 1: Digital Technology versus Information Technology versus Information and Communication Technology

Digital technology is used as an umbrella term for a variety of technologies. This includes subsets such as information technology (IT) and information and communication technology (ICT). IT refers to computing technologies including hardware, software, the internet, and others.[a] Similar to IT, ICT refers to technologies that provide access to information through telecommunication, with a focus on communication. Typical ICT includes the internet, mobile phones, and wireless networks.[b] One can see that the definitions for IT and ICT overlap thematically. However, both are subsets of digital technologies which include electronic devices, systems, and resources that generate, store, or process data and information.[c]

[a]   P. Christensson. 2006. *IT Definition*. https://techterms.com/definition/it.
[b]   P. Christensson. 2006. *ICT Definition*. https://techterms.com/definition/ict.
[c]   Victoria State Government Education and Training. 2019. *Teach with digital technologies*. https://www.education.vic.gov.au/school/teachers/teachingresources/digital/Pages/teach.aspx.

Introduction of Various Existing and Emerging Digital Technologies and Their Respective Applications in Various Fields

There are various potential ways of dividing or categorizing the development of DT into stages or levels of development, including discussions of their role in the third and fourth industrial revolutions. A number of publications have attempted to categorize such technologies. Based on a review of relevant articles[3] and information technology databases that attempt such structuring,[4] this publication proposes to use the framework of Stages I, II, and III below, laying out the attributes of each stage, and the relevant actions it enables.

On this basis, this publication places DTs on the Gartner Hype Cycle—an approach developed by research and advisory firm, Gartner, to discern the hype around new technologies from their commercial viability.[5] The technologies were not assigned to points on the graph. Rather, they were placed into sections of the Gartner Hype Cycle as each technology only represents the main one of several subcategories. The technologies were then classified into rough sections, as shown in Figure 3.

DTs of Stage I are currently being used for a range of supporting actions relevant to CCM, CCA, DRM, and environmental sustainability. These technologies, such as mobile phone networks[6] and satellite imagery,[7] have already experienced their commercial breakthrough. DTs of Stage I have become an integral part of today's world and to a certain extent have prepared the way for DTs of the other two stages. Therefore, Stage I technologies are all found on the Plateau of Productivity, the last parts of Gartner Hype Cycle, meaning mainstream adoption took off (footnote 5).

---

[3]   T. Haigh. 2011. The history of information technology. *Annual Review of Information Science and Technology*, 45 (1). Also: S. Berger, M. Denner, and M. Roeglinger, Maximilian. 2018. The Nature of Digital Technologies - Development of a Multi-Layer Taxonomy. *Research Papers*. 92. https://aisel.aisnet.org/ecis2018_rp/92.
[4]   WIPO GREEN. 2020. Climate-Friendly Information and Communication Technology. 17 May. https://www3.wipo.int/wipogreen/en/news/2020/news_0021.html.
[5]   Gartner. n.d. Gartner Hype Circle. https://www.gartner.com/en/research/methodologies/gartner-hype-cycle (accessed 25 November 2020).
[6]   P. Staunstrup n.d. *Breakthrough for mobile telephony*. https://www.ericsson.com/en/about-us/history/products/mobile-telephony/breakthrough-for-mobile-telephony (accessed 25 November 2020).
[7]   NASA Kennedy Space Center Visitor Complex. 2019. 60 Years Ago First Satellite Image of Earth. https://www.kennedyspacecenter.com/blog/60-years-ago-first-satellite-image-of-earth (accessed 25 November 2020).

Figure 3: Digital Technologies on the Gartner Hype Cycle

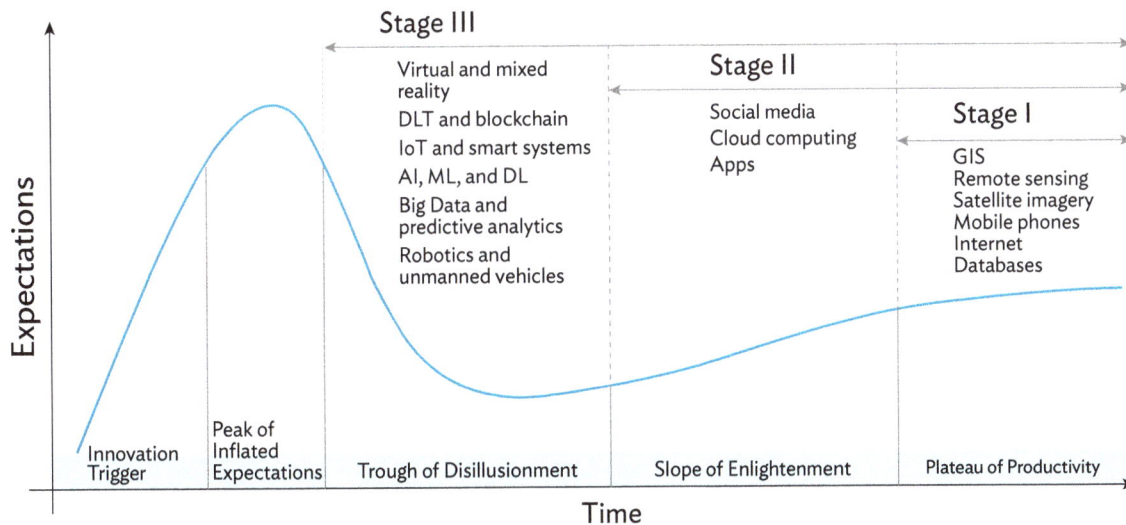

Stage III
- Virtual and mixed reality
- DLT and blockchain
- IoT and smart systems
- AI, ML, and DL
- Big Data and predictive analytics
- Robotics and unmanned vehicles

Stage II
- Social media
- Cloud computing
- Apps

Stage I
- GIS
- Remote sensing
- Satellite imagery
- Mobile phones
- Internet
- Databases

Expectations / Time

Innovation Trigger — Peak of Inflated Expectations — Trough of Disillusionment — Slope of Enlightenment — Plateau of Productivity

AI = artificial intelligence, DL = deep learning DLT = distributed ledger technology, GIS = geographic information system, IoT = Internet of Things, ML = machine learning.

Source: Asian Development Bank using the Gartner Hype Cycle available at Gartner. 2019. *Gartner 2019 Hype Cycle Shows Most Blockchain Technologies Are Still Five to 10 Years Away From Transformational Impact.* https://www.gartner.com/en/newsroom/press-releases/2019-10-08-gartner-2019-hype-cycle-shows-most-blockchain-technologies-are-still-five-to-10-years-away-from-transformational-impact.

Stage II DTs are available, but are not being utilized to their full potential to address CCM, CCA, DRM, and environmental sustainability. All technologies from Stage II are built upon Stage I technologies. There would be no apps, social media, or cloud computing as we know them without the internet or mobile phones. Commercial breakthroughs of these technologies occurred rather recently. Facebook only reached 100 million users in 2008, which is negligible compared to the approximately 2 billion users in 2017,[8] and one of the largest app stores, the Google Play Store, only launched in 2012.[9] Compared to DTs from Stage I, this is rather recent. Stage II technologies are also found at the end of the Hype Cycle, as mainstream adoption already started taking off. Nevertheless, these technologies spread also to the Slope of Enlightenment of the cycle as more instances have to learn how such technologies can benefit their endeavors. In 2019, only one out of four businesses used cloud computing. According to a survey by the European Commission, some obstacles, for example, lack of trust due to security risks, have to be overcome for larger-scale implementation of cloud computing.[10]

Technologies of Stage III are expected to reach their commercial breakthroughs in the future and have the potential to significantly accelerate actions to address CCM, CCA, DRM, and environmental sustainability. The managing consulting company PricewaterhouseCoopers expects DTs of this stage to yield the next wave of innovation.[11] A survey, conducted in 2020, of executives in organizations with ongoing Artificial Intelligence-initiatives and annual sales of more than $1 billion found that 47% had launched AI pilot programs.[12] That shows that enterprises

[8]   M. Rouse. 2019. *Google Play (Android Market)*. https://searchmobilecomputing.techtarget.com/definition/Google-Play-Android-Market#:~:text=The%20Google%20Play%20Store%20launched,over%20190%20countries%20and%20territories (accessed 25 November 2020).

[9]   E. Ortiz-Espina. 2019. *The rise of social media*. https://ourworldindata.org/rise-of-social-media (accessed 25 November 2020).

[10] European Commission. 2019. *Cloud Computing – Brochure*. https://ec.europa.eu/digital-single-market/en/news/cloud-computing-brochure (accessed 25 November 2020).

[11] PwC. n.d. The Essential Eight. https://www.pwc.com/gx/en/issues/technology/essential-eight-technologies.html (accessed 25 November 2020).

[12] A. Holst. 2020. AI Implementation in Organizations Worldwide. *Statista*. https://www.statista.com/statistics/1133015/statements-best-describes-ai-implementation-in-organizations/ (accessed 25 November 2020).

understand that they can benefit from AI. The technology is allocated in the Slope of Enlightenment as only 13% of mentioned enterprises have successfully deployed used cases in their manufacturing facilities (footnote 12). Mainstream adoption has not yet fully happened. Blockchain technologies are placed in the Trough of Disillusionment, an earlier phase of the cycle. According to Gartner in 2019, blockchain technologies still have to prove themselves as many enterprise blockchain projects are still in an experimentation phase.[13]

Stage I: Internet, Satellite Imagery, Geographic Information System, Remote Sensing, Mobile Phones, and Databases

Stage I technologies have been in existence for several decades, and provide basic information, communication, and search capabilities to users.

The internet consists of a global network of computers and other electronic devices. These devices communicate with each other by exchanging data and information. Connecting a capable electronic device to the internet gives its user access to websites and communication platforms. Access to the internet occurs for most users through a type of software called a web browser, which serves as a search device to search for specific information or websites.[14] Since its publication in the 1990s, the internet's user base steadily grew up to 4.57 billion in 2019.[15]

Table 1 shows that the number of individuals using the internet is growing among the ADB members considered. The percentage of the population which uses the internet is strongly related to the average income of the respective country.[16] That percentage grows with the economic income of the country. Low-income economies, such as Afghanistan, have lower internet usage among its population, compared to high-income economies such as the Republic of Korea.[17]

As Table 1 shows, only half of the world is connected to the internet, and the percentage of internet users in developed economies is significantly higher than in developing countries. On the one hand, this shows that only half of the world's population has access to the digital economy and can make use of internet applications that help to adapt and mitigate climate change, improve environmental sustainability or prepare for disasters. On the other hand, it appears that a simple solution to connecting such economies to the digital economy and other benefits of the internet is to give them access to this technology to increase the number of internet users. BRCK, a technology company from Kenya, for example, is specialized and successful in connecting frontier markets with the internet, which is what they are currently doing in East Africa.[18]

Next, consider **satellite imagery, geographic information systems (GISs), and remote sensing.** Satellites are best known as technical devices that orbit the earth. They may contain camera systems and sensors which serve different purposes. The bird's-eye view from space allows the satellites to explore larger areas than what would be possible from within the Earth's atmosphere.[19] To help scientists predict weather and climate, satellites provide information about clouds, oceans, land, and ice. All this helps to better understand the atmospheric

---

[13]   Gartner. 2019. *Gartner 2019 Hype Cycle Shows Most Blockchain Technologies Are Still Five to 10 Years Away From Transformational Impact.* https://www.gartner.com/en/newsroom/press-releases/2019-10-08-gartner-2019-hype-cycle-shows-most-blockchain-technologies-are-still-five-to-10-years-away-from-transformational-impact.

[14]   GCF Global. 2020. *Internet Basics - What is the Internet?* https://edu.gcfglobal.org/en/internetbasics/what-is-the-internet/1/.

[15]   R. Shuler. 2002. How does the Internet work? *Stanford University.* https://web.stanford.edu/class/msande91si/www-spr04/readings/week1/InternetWhitepaper.htm.

[16]   Pew Research Center. 2015. *Internet Access Strongly Related to Per Capita Income.* https://www.pewresearch.org/global/interactives/internet-usage/.

[17]   World Bank. n.d. *Individuals using the Internet (% of population).* https://data.worldbank.org/indicator/IT.NET.USER.ZS?end=1995&locations=CN-KR-PH-TH-BD-IN-ID-AF-KH-MN-VU&name_desc=false&start=1995&view=bar (accessed 25 November 2020).

[18]   BRCK. 2020. Connecting Africa to the internet. BRCK. https://www.brck.com/.

[19]   D. Stillman. 2014. *What Is a Satellite?* https://www.nasa.gov/audience/forstudents/5-8/features/nasa-knows/what-is-a-satellite-58.html (accessed 25 November 2020).

Table 1: Individuals Using the Internet among Few ADB Member Countries

	Individuals Using the Internet (% of population)									
Country	2000	2002	2004	2006	2008	2010	2012	2014	(2015)/2016	(2017)/2018
Afghanistan	< 0.1	< 0.1	0.1	2.1	1.8	4.0	5.5	7.0	8.3 (2015)	11.5 (2017)
India	0.5	1.5	2.0	2.8	4.4	7.5	12.6	21.0	22.0	20.1
People's Republic of China	1.8	4.6	7.3	10.5	22.6	34.3	42.3	47.9	53.2	54.3 (2017)
Republic of Korea	44.7	59.4	72.7	78.1	81.0	83.7	84.1	87.6	92.8	96.0
World	6.7	10.5	14.1	17.5	23.0	28.8	34.2	39.9	44.8	49 (2017)

Source: World Bank. 2019. Individuals using the Internet (% of population). https://data.worldbank.org/indicator/IT.NET.USER.ZS?end=1995&locations=CN-KR-PH-TH-BD-IN-ID-AF-KH-MN-VU&name_desc=false&start=1995&view=bar.

state of the world and, among other things, provides the foundation of measurements concerning the patterns of global environmental and climatic changes.[20]

As of mid-2020, about 2,200 satellites orbited the earth.[21] These included different types such as earth observation satellites, communications satellites, navigation satellites, and weather satellites.[22] In mid-2019, 768 of such satellites could produce images.[23]

The process of collecting such information remotely is called remote sensing. It refers to sensors attached to satellites or aircraft to collect certain information either passively, by external stimuli as natural energy reflected from the earth; or actively, by projecting a laser beam onto the earth's surface to scan its topography.[24]

Satellite imageries resulting from the recordings have become an essential tool to monitor land-use changes caused by socioeconomic factors.[25] Computer systems like GIS that process data and pictures provided by satellites, for example, through remote sensing, can be used to separate imageries into different layers such as streets, buildings, and vegetation. Separating these layers helps to identify connections that would otherwise have remained hidden. This can be used for disaster prevention when it comes to identifying places that are likely to be exposed to natural hazards[26] and has been widely applied in Asia.[27]

Mobile phones provide the basic functions of mobile reachability through text messaging and phone calls for its user (smartphones are considered part of Stage II technologies in the context of this report and, therefore, discussed in

---

[20] National Geographic. n.d. *GIS (Geographic Information System)*. https://www.nationalgeographic.org/encyclopedia/geographic-information-system-gis/12th-grade/ (accessed 25 November 2020); ScienceDirect. n.d. *Satellite Imagery*. https://www.sciencedirect.com/topics/earth-and-planetary-sciences/satellite-imagery (accessed 25 November 2020).

[21] *Pittsburgh Post Gazette*. 2020. A crowd in space: Tens of thousands of satellites planned for orbit. 20 May. https://www.post-gazette.com/opinion/editorials/2020/05/20/space-satellites-crowded-junk-exploration-impact/stories/202002190057.

[22] G. Petelin, M. Antoniou, and G. Papa. 2021. Multi-objective approaches to ground station scheduling for optimization of communication with satellites. Optimization and Engineering, pp. 1–38.

[23] C. Beam. 2019. Soon satellites will be able to watch you everywhere all the time. MIT Review. 26 June 2019. https://www.technologyreview.com/2019/06/26/102931/satellites-threaten-privacy/.

[24] NOAA. 2020. *What is remote sensing?* https://oceanservice.noaa.gov/facts/remotesensing.html (accessed 25 November 2020).

[25] PhD Essay. 2020. *Socio-Economic and Environmental Impacts of Land Use Change*. https://phdessay.com/socio-economic-and-environmental-impacts-of-land-use-change/ (accessed 25 November 2020).

[26] Statista. n.d. Number of smartphone users from 2016 to 2021. https://www.statista.com/statistics/330695/number-of-smartphone-users-worldwide/ (accessed 25 November 2020).

[27] K. Wang. 2015. *Utilizing space and GIS for effective disaster risk management -ESCAP's practices in Asia and the Pacific*. http://www.unoosa.org/pdf/pres/copuos2015/copuos2015tech21E.pdf (accessed 25 November 2020).

that section). **Feature phones** are phones that are classified between simple mobile phones and smartphones. They may still use buttons rather than touchscreens, but are able to play music and connect to the internet. However, their other functions are very limited compared to smartphones because they use operating systems with limited memory.[28] Feature phones have the advantages of being easy to operate, having long battery runtimes, and being relatively cheap. These advantages have enabled over 400 million people to use feature phones in India in 2018.[29]

A major use of mobile and feature phones could be in promoting citizen science. For many decades, citizen science, in which members of the general public collect and analyze data related to the environment, often in collaboration with scientists, has been used in environmental monitoring.[30]

A **database** is a comprehensive, sometimes exhaustive, collection of files and/or records pertaining to a specific subject.[31] Naturally, DTs have had a dramatic effect on the main services of databases, including collection, cleaning, filtering, integration, sharing, storage, transfer, visualization, analysis, and security. DTs have also enabled progress in the searchability of databases, including optimization and personalization of the search process.[32]

Stage II: Social Media, Apps, Smartphones, and Cloud Computing

Stage II technologies have been in existence for a number of years and, in the context of this report, cover instant communication and collaboration, and the instant sharing and storing of documents.

Social media are platforms and blogs to create, share, and/or exchange information and ideas in virtual communities and networks.[33] While originally, many social media platforms came from the United States (US) (Facebook, Instagram, Twitter, YouTube), Asian platforms have now risen to huge membership bases (notably WeChat with over 1.1 billion users as of the third quarter of 2020), and have developed specific concepts that have allowed them to dominate a large part of the social media market in Asia (e.g., Line, by offering a complete digital ecosystem of messaging, payments, online shopping, and other features; or Kakao Talk, by combining standard information exchange via messages, photos, user location with emoticons, and sticker collections that have made it hugely popular in countries such as the Republic of Korea). Social media has become an important part of internet technologies and communities. Out of 4.57 billion internet users, 3.96 billion were active users of social media in July 2020.[34] In the regional context, all top three geographic areas with the most users globally were in Asia as of July 2020 (East Asia, Southeast Asia, and South Asia) (footnote 34).

Social media can make various contributions to solving environmental challenges. Companies from different industries may use social media to communicate their sustainability strategy, and customers are increasingly able to challenge the interpretation of the companies' sustainability record.[35] In addition, social media also allows

---

28   Energizer. n.d. *KaiOS Operating System.* https://www.energizeyourdevice.com/en/mobiles/product/details/kaios-operating-system/ (accessed 25 November 2020).

29   Y. Bhatia. 2018. *Sudden escalation of the feature phone market in India.* https://telecom.economictimes.indiatimes.com/tele-talk/sudden-escalation-of-the-feature-phone-market-in-india/2984 (accessed 25 November 2020).

30   K. Albus, R. Thompson and F. Mitchell. 2019. *Usability of Existing Volunteer Water Monitoring Data: What Can the Literature Tell Us?* Citizen Science: Theory and Practice, 4 (1). https://theoryandpractice.citizenscienceassociation.org/articles/10.5334/cstp.222/.

31   OECD Glossary of statistical terms. https://stats.oecd.org/glossary/.

32   L. Bellatreche, P. Valduriez, and T. Morzy. 2018. Advances in Databases and Information Systems. Inf Syst Front 20, 1–6.

33   Investopedia. n.d. *Social Media.* https://www.investopedia.com/terms/s/social-media.asp (accessed 18 April 2021)

34   Statista. n.d. *Number of social network users worldwide in 2020, by region.* https://www.statista.com/statistics/454772/number-social-media-user-worldwide-region/ (accessed 25 November 2020).

35   Almost any sustainability issue created by companies is widely discussed on social media, and the corporate narrative challenged by individual users. Large-scale negative environmental impacts are frequently highlighted around the world in this way, and may trigger corporate action. One of the most recent prominent examples is the increasing worldwide concern among social media users with ocean plastic pollution (to only mention two sites rallying social media users to get involved: https://oceana.org/our-campaigns/plastics and https://oneworldoneocean.com/get-involved/). The world's largest plastics producers have since put together response strategies, including funds such as the Closed Loop Venture and Infrastructure funds to tackle the inflow of plastics into the ocean from the key ,plastics source countries.

for data scraping, which can be used in the climate change, sustainability, and DRM arenas, for example, by measuring sustainability reporting performance using web scraping.[36]

Social media facilitates people to express views and exchange information on issues of importance to them, including the dynamic and complex systems underlying climate change, DRM, or environmental sustainability issues.[37] Among the issue areas discussed here, this has been particularly the case for environmental quality issues affecting human health. Blogging activities via Weibo or WeChat in the People's Republic of China (PRC) have proven to be a reliable source for determining air quality and particulate matter.[38] Also, social media activities can help to draw conclusions on human behavior in the event of disasters and to model escape routes.[39] By localizing these blog entries, disaster relief can be provided quickly and efficiently.[40]

Applications, more known as **apps**, are software programs that allow its user to perform a specific task. A distinction is made between mobile and desktop apps.[41] This publication deals with mobile apps, as these usually also cover the application areas of desktop apps. Apps span a vast range of climate, DRM, and environmental topics, with various examples given in sections 3.1–3.4. One way to indicate the types of uses they can be put to in the case of climate change is to give examples of scientifically focused apps (e.g., the US National Aeronautics and Space Administration's (NASA) Earth Now, which visualized recent global climate data, pulling information from satellites); the UN Climate Change app of the United Nations Framework Convention on Climate Change (UNFCCC) that provides updated information on the UN climate change process so that users can participate in UNFCCC events from their phones; and various apps allowing users to improve their carbon footprint in the way they commute (Commute Greener! app) or shop (Climate Counts app).

In addition to contributions to sustainable development goals and climate change adaptation and mitigation, apps can help in the event of disasters. Programs that use GIS help to make climate and weather data understandable even for lay persons and to provide early warnings of possible extreme weather events.[42] In case of a disaster, apps can inform their users with first aid instructions and keep them and also the authorities up to date on the current situation, even in real-time (footnote 40). Hence, authorities can send help quickly.

Mobile apps can offer their users a variety of services and are mostly enabled by **smartphones**, which the International Telecommunication Unit defines as mobile phones with a touchscreen display that enables its users to access advanced internet-based services and performs similar tasks like a computer.[43] For this, as with computers, smartphones use an operating system and can run various applications.[44] It is estimated that about 6.4 billion people will be using smartphones by 2021.[45]

---

36   UK Office for National Statistics. Measuring Sustainability Reporting Using Web Scraping and Natural Language Processing. (no date)

37   This report defines "systems" here as social, economic, or ecological systems—or "any dynamic systems characterized by interdependence, mutual interaction, information feedback, and circular causality" (quote from the System Dynamics Society, systemdynamics.org).

38   W. Jiang et al. 2015. Using social media to detect outdoor air pollution and monitor air quality index (AQI): a geo-targeted spatiotemporal analysis framework with Sina Weibo (Chinese Twitter). PloS one, 10(10), p.e0141185.

39   R. Ilieva. 2018. Social Media Data for Urban Sustainability. https://sustainabilitycommunity.springernature.com/posts/39900-social-media-data-for-urban-sustainability (accessed 25 November 2020).

40   Ushahidi. https://www.ushahidi.com/.

41   GCF Global. 2020. What is an Application? https://edu.gcfglobal.org/en/computerbasics/understanding-applications/1/ (accessed 25 November 2020).

42   J. Rotter and M. Price. 2019. The best emergency apps for wildfires, earthquakes, and other disasters. CNET. https://www.cnet.com/news/the-best-emergency-apps-for-wildfires-hurricanes-earthquakes-and-other-disasters/.

43   ITU. 2017. ITU Expert group on household indicators (EGH). Background document 3. Proposal for a definition of Smartphone. Geneva. https://www.itu.int/en/ITU-D/Statistics/Documents/events/egh2017/EGH%202012%20background%20document%203%20-%20Definition%20of%20smartphone.pdf, https://de.statista.com/themen/581/smartphones/.

44   Tufts University. 2020. Social Media Overview. https://communications.tufts.edu/marketing-and-branding/social-media-overview/; Statista. 2020.Global digital population as of October 2020. https://www.statista.com/statistics/617136/digital-population-worldwide/.

45   Statista. 2021. Number of smartphone users worldwide from 2016 to 2023. https://www.statista.com/statistics/330695/number-of-smartphone-users-worldwide/ (accessed 15 September 2021).

Similar to the number of internet users, the penetration rate of smartphones as a percentage of the population depends on the income of the country, as shown in Table 2. Although no data is available on smartphone penetration in Afghanistan, the number of cell phone subscriptions of around 58 per 100 people in 2019 compared to 35 per 100 people in 2010[46] shows that there is a growing demand for mobile reachability. The smartphone penetration rate is expected to grow until it reaches close to 100%, as can be seen from the smartphone penetration rate of the Republic of Korea.

Table 2: Smartphone Penetration Rate as Share of the Population (%)

Country	2017	2018	2019	2020	2021	2022	2023	2024	2025
India[a]	26	29	31	32	34	34	35	-	-
PRC[b]	48	50	53	55	57	59	61	-	-
Republic of Korea[c]	89.8	91.8	93.6	95.4	97.1	97.4	97.4	97.4	97.4
World[d]	33.8	36	38.2	41.6	44.9	48.1	-	-	-

PRC = People's Republic of China.

[a] Statista. n.d. *Mobile phone internet user penetration India 2015–2023.* https://www.statista.com/statistics/309019/india-mobile-phone-internet-user-penetration/ (accessed 25 November 2020).

[b] Statista. n.d. *Smartphone penetration rate in China from 2015 to 2023.* https://www.statista.com/statistics/321482/smartphone-user-penetration-in-china/ (accessed 25 November 2020).

[c] Statista. n.d. *Smartphone penetration rate as share of the population in South Korea from 2015 to 2025.* https://www.statista.com/statistics/321408/smartphone-user-penetration-in-south-korea/ (accessed 25 November 2020).

[d] Statista. n.d. *Number of smartphone users from 2016 to 2021.* https://www.statista.com/statistics/330695/number-of-smartphone-users-worldwide/ (accessed 25 November 2020); World Bank. n.d. Population, total. https://data.worldbank.org/indicator/SP.POP.TOTL?end=2019&start=2015 (accessed 25 November 2020); Worldometer. n.d. World Population Projections. https://www.worldometers.info/world-population/world-population-projections/ (accessed 25 November 2020).

Cloud computing is a service accessible via the internet and serves companies to access computing power, data storage, and databases (footnote 40). This technology stands for a service that provides convenient and on-demand access to a shared pool of computer resources which then can be used with minimal administrative effort.[47] The European Commission describes cloud computing as an essential condition for an innovative economy[48] as it expects efficiency gains in data processing through the use of cloud computing.[49] Thus, the creation of new enterprises and job opportunities will accompany the large-scale application of cloud computing. This will significantly increase the gross domestic product of the European Union.[50]

New sources of data, for example, provided by satellites, and sensors in water, on land, and in the air deliver enormous data volumes. If institutions lack the computing power to evaluate these data sets, cloud computing can be the solution without these institutions having to build their own data centers.[51] Such data can help to measure the progress on the Sustainable Development Goals and other environment-related data and benefit significantly from their integration with cloud computing.[52]

---

[46] World Bank. n.d. *Mobile cellular subscriptions (per 100 people) – Afghanistan.* https://data.worldbank.org/indicator/IT.CEL.SETS.P2?locations=AF (accessed 25 November 2020).

[47] AWS. n.d. *What is cloud computing?* https://aws.amazon.com/what-is-cloud-computing/ (accessed 25 November 2020).

[48] Federal Office for Information Security. n.d. *Cloud Computing Basics.* https://www.bsi.bund.de/EN/Topics/CloudComputing/Basics/Basics_node.html (accessed 25 November 2020).

[49] European Commission. 2020. *Cloud computing.* https://ec.europa.eu/digital-single-market/en/cloud (accessed 25 November 2020).

[50] European Commission. 2017. *Measuring the economic impact of cloud computing in Europe.* https://ec.europa.eu/digital-single-market/en/news/measuring-economic-impact-cloud-computing-europe (accessed 25 November 2020).

[51] European Commission. 2017. *Cloud computing.* https://ec.europa.eu/digital-single-market/en/cloud.

[52] United Nations. n.d. *Big Data for Sustainable Development.* https://www.un.org/en/global-issues/big-data-for-sustainable-development.

Stage III: Artificial Intelligence, Machine Learning, Deep Learning; Internet of Things and Smart Systems; Distributed Ledger Technology and Blockchain; Big Data and Predictive Analytics; Virtual and Mixed Reality; Robotics and Unmanned Vehicles

Stage III technologies are the most recent technology type discussed in this publication and are focused on replacing or even magnifying human capacities (**artificial intelligence [AI]** for the mind, robotics for physical action). They also go beyond the individual tools of the previous stages to create systems that use large amounts of individual components (smart systems) and data (predictive analytics) to enable DTs to solve increasingly complex problems.

One such Stage III technology is AI. Definitions of AI differ widely, but the term generally refers to computer systems that can perform tasks that normally require human intelligence (visual and speech recognition, learning, planning, problem-solving, decision-making, and language translation).[53] It is argued that the next generation of AI will have the same level of economic and social impact as the technological revolutions associated with the internet and mobile phones, which have fundamentally changed the way individuals and organizations communicate and interact with the world.[54]

Some have argued that AI is rapidly becoming a general-purpose technology (GPT). GPTs are defined as technologies that possess potential to generally transform a wide range of household as well as business activities.[55] GPTs such as AI also facilitate complementary innovations and bring transformations and changes in business processes.[56] This is particularly the case for applications of AI in combination with one or several of the other Stage III technologies discussed below. For instance, data obtained through IoT can then be processed via AI. The company Harvest AI, for example, builds IoT systems in high-tech greenhouses to collect that a proprietary AI system then processes to predict the harvest date of plants in that same greenhouse. As a GPT, learning curricula should be adopted to develop skills for an AI-based economy;[57] and AI has to be considered a new factor of production.[58]

Faster computing, big data (see Box 2 for a definition and explanation), and better algorithms (an instruction of different steps to complete a specific task)[59] have in recent years helped propel recent breakthroughs in AI,[60] including a broad range of applications in climate-related fields[61] with examples including clean distributed energy grids, precision agriculture, sustainable supply chains, environmental monitoring and enforcement, and enhanced weather and disaster prediction and response.[62] Further advances in AI will prove to be disruptive, resulting in new opportunities for collaboration between humans and machines (but also loss of traditional jobs).[63]

[53] D.G. Harkut and K. Kasat. 2019. Introductory Chapter: Artificial Intelligence-Challenges and Applications. In Artificial Intelligence-Scope and Limitations. IntechOpen.

[54] E. Schmidt and J. Cohen. 2015. *Technology: Inventive artificial intelligence will make all of us better.* http://time.com/4154126/technology-essay-eric-schmidt-jared-cohen.

[55] B. Jovanovic and P. Rousseau. 2005. *General Purpose Technologies.* The National Bureau of Economic Research. https://www.nber.org/papers/w11093.

[56] Gill et al. 2020. The Economics of AI-Based Technologies: A Framework and an Application to Europe. https://papers.ssrn.com/sol3/papers.cfm?abstract_id=3660114#.

[57] M. Trajtenberg. 2018. AI as the next General Purpose Technology: A Political-Economy Perspective. *National Bureau of Economic Research Working Paper Series.* 2018.

[58] Accenture: How AI can drive South America's Growth. 2017.

[59] GCF Global. 2020. *Computer Science-Algorithms.* https://edu.gcfglobal.org/en/computer-science/algorithms/1/.

[60] Enabling Digital Development – Six digital technologies to watch. https://link.springer.com/chapter/10.1007/978-981-32-9915-3_9.

[61] State of the Planet. 2018. *Artificial Intelligence—A Game Changer for Climate Change and the Environment.* https://blogs.ei.columbia.edu/2018/06/05/artificial-intelligence-climate-environment/.

[62] PwC. 2019. Using AI to better manage the environment could reduce greenhouse gas emissions, boost global GDP by up to US $5 trillion and create up to 38m jobs by 2030. https://www.pwc.com/gx/en/news-room/press-releases/2019/ai-realise-gains-environment.html.

[63] How AI can enable a sustainable future. https://www.pwc.co.uk/services/sustainability-climate-change/insights/how-ai-future-can-enable-sustainable-future.html; What AI means for Sustainability. https://www.greenbiz.com/article/what-artificial-intelligence-means-sustainability; How IoT and AI can enable Environmental Sustainability. https://www.forbes.com/sites/cognitiveworld/2019/09/04/how-iot-and-ai-can-enable-environmental-sustainability/#61e14b0268df.

Box 2: What Is Big Data?

The research company, Gartner, defines big data as "high-volume, high-velocity and high-variety information assets that demand cost-effective, innovative forms of information processing for enhanced insight and decision-making."[a] Put differently, big data involves the availability of data in real time, at larger scale, with less structure, and on different types of variables than previously used.[b]

ᵃ   Gartner. 2013. Big data. http://www.gartner.com/it-glossary/big-data/.
ᵇ   L. Einav and J. Levin. 2013. The data revolution and economic analysis, *NBER Innovation Policy and the Economy Conference*. April. http://www.nber.org/papers/w19035.

An interesting subset of AI, which can be used in combination with satellite imagery but is also applied in self-driving and modern cars, is computer vision (CV). CV tries to enable computers to mimic human vision. In modern vehicles, such technologies can be already applied, for example, to recognize road signs and either warn the driver that the speed limit is reduced or slow down the car automatically. In satellite imagery, it may help to automatically detect plastic pollution or wildfires. CV works with recognizing patterns. One way to teach such a program a certain pattern can be machine learning.[64]

Machine learning (ML) is a subset of AI and though the terms are often used interchangeably, they are not the same thing.[65] Rather than carrying out smart tasks, as AI is considered to do, ML provides systems the ability to automatically learn and improve from experience without being explicitly programmed.[66] AI is the broader concept of machines being able to intelligently carry out tasks; ML is the idea that machines just require data and can learn by themselves on that basis.

A type of ML, known as deep learning (DL), uses algorithms which teach computers to learn by examples and perform tasks based on classifying structured as well as unstructured data such as images, sound, and text. A simple way to view DL is to consider it as a neural network model involving many layers. The exponential growth of computing power allows DL models to build up neural networks with many layers, which was not possible in classical neural networks. Due to this, DL can analyze complex nonlinear patterns in high dimensional data that can no longer be represented by traditional mathematical models and where a human typically could not recognize any kind of pattern or relationship in such data. Such patterns involve a large number of traits. Traditional ML algorithms, on the other hand, work when the number of traits is small.[67]

Next, the **Internet of Things (IoT) and smart systems** refer to the interconnection of objects to internet infrastructure through embedded computing devices (e.g., sensors), often leading to the processing of such data in real-time.[68] See Box 3 for a definition and explanation and Box 4 for an example of utilization. There are generally five broad categories for technologies in this space, including wearable devices, smart homes, smart cities,

---

64   I. Mihajlovic. 2019. *Everything You Ever Wanted To Know About Computer Vision*. https://towardsdatascience.com/everything-you-ever-wanted-to-know-about-computer-vision-heres-a-look-why-it-s-so-awesome-e8a58dfb641e.

65   B. Marr. 2016. *What Is The Difference Between Artificial Intelligence And Machine Learning?* https://www.forbes.com/sites/bernardmarr/2016/12/06/what-is-the-difference-between-artificial-intelligence-and-machine-learning/#52916c542742.

66   D.D. Luxton. 2016. An Introduction to Artificial Intelligence in Behavioral and Mental Health Care. *Artificial Intelligence in Behavioral and Mental Health Care*. 2016. http://documents1.worldbank.org/curated/en/896971468194972881/310436360_201602630200216/additional/102725-PUB-Replacement-PUBLIC.pdf.

67   F. Jiang et al. 2017. Artificial intelligence in healthcare: past, present and future. *Stroke and Vascular Neurology*. https://svn.bmj.com/content/2/4/230.full.

68   Some readers will find similarities between the interconnectedness in IoT and smart systems, on the one hand, and supervisory control and data acquisition (SCADA), on the other hand. SCADA is a system that monitors and controls industrial, infrastructure, and facility-based processes over a variety of devices. While there has been some discussion about whether IoT will at least partially replace SCADA, a more common view is that IoT will tend to be implemented on top of SCADA.

environmental sensors, and business applications. The efficiency gains from this lead to an extremely broad range of potential sustainability applications of this, given that connected sensors can deliver data from various fields and for various appliances, including energy, water, agriculture (where farms can use sensors to monitor soil conditions and guide autonomous irrigation systems),[69] transport, and urban management (smart traffic synchronization systems in cities lead to savings on travel time and fuel),[70] buildings (building automation and controls, lighting and mechanical upgrades),[71] and solid waste.[72] Data that is provided by sensors in such areas can be recorded in databases and evaluated by big data processes and AI.

Challenges of IoT and smart systems include a fragmented landscape of standardization that prevent interoperability, the relatively high cost of embedded devices, privacy, and security concerns.[73]

Box 3: What Is the Internet of Things?

The Internet of Things (IoT) is the network of physical objects or "things" such as machines, devices, appliances, animals, or people that are embedded with electronics, software, and sensors and are provided with unique identifiers.[a] They possess the ability to transfer data across the internet with minimal human interventions.

According to Gartner, there are three components of an IoT service: the edge, the platform, and the user. The edge is the location where data originates or is aggregated. Data may also be reduced to the essential or minimal parts. In some cases, data may be analyzed. The data then goes to the platform, which is typically in the cloud. Analytics are often performed in the cloud using algorithms.

Data that have been analyzed move from the IoT platform to the user. For that to happen, the user deploys an application program interface (API) to call or query the data, which specifies the way software components of the user and platform interact.[b] It is also possible that the IoT finds a predetermined set of events and then announces or signals to the user.

[a]   M. Rouse. 2019. Definition-Internet of Things (IoT). techtarget. com, para. 1.
[b]   N. Laskowski. 2016. Delving into an enterprise IoT initiative? Read this first. http://searchcio.techtarget.com/feature/Delving-into-an-enterprise-IoT-initiative-Read-this-first.

---

[69]   J.D. Borrero and A. Zabalo. 2020. An autonomous wireless device for real-time monitoring of water needs. Sensors, 20(7), p. 2078.
[70]   Various writings point to the important distinctions between AI and ML. E.g., R. Iriondo. 2018. Machine Learning (ML) vs. Artificial Intelligence (AI) — Crucial Differences. *Towards AI*. 16 October. https://pub.towardsai.net/differences-between-ai-and-machine-learning-and-why-it-matters-1255b182fc6. Machine learning focuses on the development of computer programs that can access data and use it to learn for themselves. ML can notably help to (cheaply) structure the plethora of available climate change and sustainability-related data. S. Ermon. n.d. Machine Learning and Decision Making for Sustainability. https://cs.stanford.edu/~ermon/slides/ermon_ijcai_early.pdf; Gauging the precise potential and bang for buck of ML in the sustainability space may require further data in itself. B. D'Amico et al. 2019. Machine learning for sustainable structures: A call for data. *In Structures* (Vol. 19, pp. 1-4). Elsevier.
[71]   For example, Singapore uses smart networks that combine global positioning system (GPS), sensor information from monitoring cameras, and other sources of sense population movement, ease traffic congestion, and reroute traffic in case of special events and emergencies.
[72]   Increased energy efficiency by about 70 % from 2004 to 2018. Honeywell Building Solutions.
[73]   A. Salam. 2020. Internet of Things for Environmental Sustainability and Climate Change. In: Internet of Things for Sustainable Community Development. *Internet of Things (Technology, Communications and Computing)*. Springer, Cham. https://doi.org/10.1007/978-3-030-35291-2_2.

Box 4: Da Nang City Utilizes Internet of Things to Respond to Pollution and Climate Change

Viet Nam's Da Nang City, which is a major port city, has developed big data-based models to assess climate risks. Its Climate Change Coordination Office develops strategies to minimize the risks and respond to factors such as pollution and climate change.[a] The city uses the Internet of Things (IoT) to address congestion problem in transportation networks.[b] Da Nang uses big data for predicting and preventing congestion on roads, and to coordinate responses in case of adverse weather or accidents.[c] Data is aggregated from multiple sources. Sensors embedded in roads, highways, and buses detect anomalies and control traffic flow.[d] The system also gives the Department of Transport access to real-time information for its fleet of buses. From the city's traffic control center, city officials can monitor traffic and control the traffic light system. In case of a congestion due to an accident, traffic lights can be adjusted, which allows time for cars affected by the jam to pass through (footnote c). This data is accessible to passengers through video screens at bus stations or via mobile apps.[e] Users can see details such as the location of a bus, speed, and predicted time of a trip. The plan in the future is to alert citizens how crowded a bus is likely to be when it arrives. Da Nang's transport grid was established based on people's movement pattern, and the city's growth pattern (footnote c).

The fifth generation (5G) cellular technology is expected to enhance the use cases of IoT. As of August 2020, 38 countries in the world had deployed 5G networks.[f] 5G offers faster connections, lower response time, and better coverage compared to its predecessors.[g] (See Box 5 for the benefits of 5G networks.) This allows more IoT systems to be used in critical areas, whose operating principle depends on low latency. Massive improvements are expected, for example, to make traffic flow more efficiently and to control prioritized fully occupied vehicles. This will reduce emissions.[h] It is also worth noting that among the world's top five economies in terms of 5G speed as of October 2020, three were ADB regional member economies: the Republic of Korea (no. 2), Australia (no. 3), and Taipei,China (no. 4).[i]

[a] McKinsey&Company. 2018. Smart Cities in Southeast Asia. https://www.mckinsey.com/business-functions/operations/our-insights/smart-cities-in-southeast-asia.

[b] V.D. Ngo and N. Kshetri. 2017. Business Opportunities and Barriers for Big Data in Vietnam. Presentation for the 2017 PTC telecommunications conference. Hawaii. 15–18 January 2017. https://online.ptc.org/ptc17/program-and-attendees/proceeding.html?pid=242.

[c] J. Dubow. 2014. The World Bank Big data and urban mobility. Cairo. 2 June 2014.

[d] P. Mallya. 2014. Big Data: The invisible force reshaping our world. 15 March. http://www.thestar.com.my/Tech/Tech-Opinion/2014/03/15/Big-Data-The-invisible-force-reshaping-our-world/.

[e] M. Wheatley. 2013. Vietnam's Cities Use Big Data To Ward Off Traffic & Pollution. http://siliconangle.com/blog/2013/08/16/vietnams-cities-use-big-data-to-ward-off-traffic-pollution/.

[f] K. Buccholz. 2020. Where 5G Technology Has Been Deployed. Statista. https://www.statista.com/chart/23194/5g-networks-deployment-world-map/.

[g] P. Collela. n.d. 5G and IoT: Ushering in a new era. https://www.ericsson.com/en/about-us/company-facts/ericsson-worldwide/india/authored-articles/5g-and-iot-ushering-in-a-new-era.

[h] T-Mobile for Business. 2019. How The 5G Era Could Help Build A More Sustainable Future. Forbes. https://www.forbes.com/sites/tmobile/2019/10/21/how-the-5g-era-could-help-build-a-more-sustainable-future/.

[i] Zee Media Bureau. 2020. Saudi Arabia has fastest 5G download speed, S Korea second --Full list of 15 countries. https://zeenews.india.com/technology/saudi-arabia-has-fastest-5g-download-speed-s-korea-second-full-list-of-15-countries-2318863.html.

Next, **distributed ledger technology (DLT), including blockchain**, hold a different set of promises for climate change and sustainability.[74] (Other types of DLT, which this publication will not go into, include Directed Acyclic Graph or Hashgraph.) A DLT represents a consensus of replicated, shared, and synchronized digital data geographically spread across multiple locations, and with no central administrator.[75] It is now recognized as a GPT for a wide range of economic activities, as long as these activities are based on the consensus of a database of transaction records.

---

[74]   World Bank. 2016. World Development Report 2016: Digital Dividends. Washington, DC: World Bank.

[75]   The Rodman Law Group. Distributed Ledger Technology. https://therodmanlawgroup.com/distributed-ledger-technology/ (accessed 18 April 2021)

Box 5: Benefits of Fifth Generation Networks

Fifth generation (5G) cellular networks have several attractive features which include lower latency, higher bandwidth, and a higher degree of reliability. 5G networks utilize millimeter (mm) wave, which is normally considered to be the range of wavelengths from 10 mm to 1 mm.[a] Compared to the frequency spectrum used by previous wireless network technologies, the amount of bandwidth available at millimeter wave frequencies is significantly higher. Such frequencies thus provide the massive bandwidth. This is especially important to transmit data created by a large number of Internet of Things devices.

The fourth generation (4G) and long-term evolution (LTE) networks have relatively high latency, which varies between 30 milliseconds (ms) and 100 ms.[b] 5G networks get closer to the user in terms of antennas, which reduces the latency. This feature is especially valuable and useful for services such as autonomous vehicles.

[a] Engineering and Technology History Wiki. n.d. *Millimeter Waves.* https://ethw.org/Millimeter_Waves.
[b] F. Weiller. 2018. *Breaking The Psychological Barrier To Autonomous Vehicle Adoption.* https://www.eeworldonline.com/breaking-the-psychological-barrier-to-autonomous-vehicle-adoption/.

DLT have particular promise in the context of transparency about individual and collective action that is so central to many climate and sustainability initiatives.[76] In particular, DLTs enable the protection and preservation of the quality of information and associated transparency. This is the type of role required for monitoring, reporting, and verification (MRV) systems that can be applied in emission reduction, but also when building sustainable supply chains starting from the extraction of commodities and their responsible procurement[77] to their further processing to semi-products until the completion of the product, meaning tracking and monitoring whether quality and sustainability standards have been met along the supply chain.

Several blockchain-based supply chain traceability projects have been built, which mainly rely on private, permissioned blockchain. As an example, the Responsible Sourcing Blockchain Network (RSBN), which is an industry collaboration that aims to support sustainable and responsible sourcing and production practices, has developed a blockchain platform that is built on Hyperledger Fabric.[78] RSBN members include IBM, Ford, Volkswagen, Huayou Cobalt, Volvo, Fiat Chrysler and British-Swiss commodities trading company Glencore (footnote 78). A key benefit of private blockchain projects such as the one implemented by RSBN is that they do not rely on proof-of-work consensus mechanism or "mining" such as those used in completely decentralized blockchain systems such as Bitcoin. Therefore, they do not suffer from problems such as high energy consumption and longer processing time. Ensuring the sustainability and limiting environmental impacts of DLT including blockchain is imperative.

Big data and predictive analytics are processes to analyze and systematically extract information data sets considered to be large by traditional data-processing application software, and apply advanced data analytics methods to them such as for predictive purposes. One example of its application is the development of systems by a multinational corporation to examine its environmental impact along their value chains. By developing an

---

[76] S. Davidson, P. De Filippi and J. Potts. 2016. Disrupting Governance: The New Institutional Economics of Distributed Ledger Technology. https://papers.ssrn.com/sol3/papers.cfm?abstract_id=2811995; Blockchain and Sustainability. https://blockchainhub.net/blog/blog/blockchain-sustainability-programming-a-sustainable-world/; Blockchain can be a vital tool to boost sustainability. https://www.sustainability-times.com/sustainable-business/blockchain-can-be-a-vital-tool-to-boost-sustainability/; Delivering blockchain's potential for environmental sustainability. https://www.odi.org/sites/odi.org.uk/files/resource-documents/12439.pdf.
[77] S. van den Brink and J. Huisman. Approaches to responsible sourcing in mineral supply chains. *Resources, Conservation and Recycling.* Vol 145. June 2019. pp. 389–398.
[78] N. Kshetri. 2021. Blockchain and Supply Chain Management. Elsevier. https://www.elsevier.com/books/blockchain-and-supply-chain-management/kshetri/978-0-323-89934-5.

online platform, these companies encourage their suppliers to report on their compliance with sustainability guidelines. This leads to reliable monitoring of individual members and the overall environmental impact.[79]

Virtual reality (VR) and mixed reality (MR, also augmented reality [AR] and cross reality [XR]) entail the merging of real and virtual worlds to produce new environments and visualizations, where physical and digital objects coexist and interact in real-time.[80] For instance, it can help city planners to test their plans in a virtual avatar of the city, as Autodesk (an architecture and construction software company) has done for San Francisco. In this way, city planners can, for example, develop a better understanding of what needs to be built where and how the original plans need to be adapted to be implemented more efficiently.[81]

Digital twin, which is "a virtual representation of an object, a service process, a product, or anything else that can be digitized,"[82] holds great promise. For a physical entity such as a city, a digital twin is an exact digital replica or representation of the entity (physical twin) and includes the "properties, condition, and behavior of the real-life object through models and data."[83] The digital twin possesses the capability to simulate the actual behavior of the physical twin in the deployed environment (footnote 78). Digital twins thus give a real-time view of what is happening with physical assets such as equipment (footnote 78). To give an example, Singapore has its digital twin. The government department, National Research Foundation (NRF), has created a Virtual Singapore to offer a three-dimensional (3D) semantic model of the city.[84] In such a model, the data's meaning can be related to the real world, displaying land attributes or the characteristics of different forms of transport, or the components of buildings and infrastructures (footnote 84).

Robotics and unmanned land, sea, and air vehicles (UAVs) can be used to monitor and map landscapes. However, this set of technologies has not made its way into many developing countries due in part to high up-front investments and the need for complex scientific clusters to give rise to such technologies. Some countries also limit the use of these technologies for espionage and privacy reasons. Rules need to be in place first, for example, where the UAVs are allowed to fly—not near airports, not in residential areas, and other rules.

Hyperautomation takes robotics, specifically robotic process automation (RPA) one step further. This term was introduced by the research company, Gartner. It is similar to what International Data Corporation (IDC) refers to as "intelligent process automation." The US market research company, Forrester, calls this process "digital process automation." RPA performs well in automating predefined steps in which rules define where relevant data can be located on each type of predefined device, applications, and other sources.[85] However, RPA may not deliver the automation goals if data from different sources are in different formats with different contents (footnote 85). Hyperautomation utilizes ML and machine vision to extract relevant information from diverse sources and take action automatically.

79 United Nations ECLAC and EuroClima. 2014. *Big data and open data assustainability tools: A working paper prepared by the Economic Commission for Latin America and the Caribbean*. Santiago, Chile. https://repositorio.cepal.org/bitstream/handle/11362/37158/1/S1420677_en.pdf.

80 A. Datta. 2019. What is the difference between Virtual Reality, Augmented Reality and Mixed Reality. *GeoSpatial World*. 5 September. https://www.geospatialworld.net/article/difference-virtual-reality-augmented-reality-mixed-reality/.

81 J. Naveen. 2019. *Applications Of Immersive Technologies In Smart Cities*. https://www.forbes.com/sites/cognitiveworld/2019/08/05/applications-of-immersive-technologies-in-smart-cities/.

82 C. Miskinis. 2018. What separates digital twin based simulations vs. a reality that is augmented. Challenge Advisory, https://www.challenge.org/insights/digital-twin-vs-augmented-reality/.

83 S. Haag and R. Anderl. 2018. Digital twin–Proof of concept. Manufacturing Letters, 15, pp. 64–66.

84 P. Liceras. 2019. Singapore experiments with its digital twin to improve city life. Tomorrow City. https://www.smartcitylab.com/blog/digital-transformation/singapore-experiments-with-its-digital-twin-to-improve-city-life/.

85 K. Mathias. 2020. Robotic Process Automation – pragmatic solution or dangerous illusion?. *BTOES Insights*. 10 February. https://insights.btoes.com/risks-robotic-process-automation-pragmatic-solution-or-dangerous-illusion-1-1.

3 The Role of Digital Technology

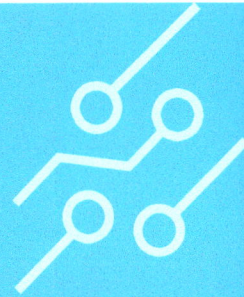

This chapter moves from the explanation of technologies at each of Stages I–III to a deep-dive into DT's role in the issue areas at the heart of OP3. Thus, it discusses conceptually and with examples the role DT at each of the three stages can play for CCM, CCA, DRM, and environmental sustainability. For each, it also lays out the problems with the DT at that stage. In Table 3 below, an overview of the respective applications of Stages I-III is provided.

Table 3: Three Stages of Digital Technologies and Their Respective Applications in the Climate Change Mitigation, Climate Change Adaptation, Disaster Risk Management, and Environmental Sustainability Arenas

Digital Technologies	Climate Change Mitigation	Climate Change Adaptation	Disaster Risk Management	Environmental Sustainability
Stage I				
Mobile phone	Communication enabler for mitigation campaigns among users.	Tracking of population migration patterns in areas affected by climate change (e.g., in Bangladesh).[a]	Mobile text alerts to communicate emergency information.[b]	Communication enablers to organize volunteer groups for environmental clean-ups.
Internet	Access to and exchange of information and data on:			
	Methods for homeowners to become energy-efficient.	Drought-resistant crops and methods for farmers.	Assessing the flood danger in a certain area.	Local options for household waste management (e.g., specific nearby recycling points).
Database	Online repository for:			
	Energy consumption in a certain part of town (completed by, for example, building types, use, and daytime).	Crop failure in past years (completed by, for example, species, region, climate data, irrigation method, fertilization method).	Climate data in and around vulnerable regions for tropical storms.	Weight and volume of food packaging (completed by type of food, calorie content, cost, place of purchase).
Satellite imagery	Assessment of emission sources (e.g., wildfires).	Assessment of land-use change due to climate change, as basis for adaptation options assessment.	Post-disaster assessment (e.g., degree of destruction, scope of the required auxiliary services, condition of remaining infrastructure).	Detection of plastic pollution in the ocean.

continued on next page

Table 3 *continued*

Digital Technologies	Climate Change Mitigation	Climate Change Adaptation	Disaster Risk Management	Environmental Sustainability
Geographic information system	Promoting resource efficiency (e.g., transport planning for low carbon cities).	Crop modeling and environmental matching procedures.	Hazard mapping of vulnerable regions (type of hazard, escape routes, available infrastructure).	Promoting environmental health and safety (e.g., appropriateness of safe waste disposal sites).
Remote sensing	Detection of gas leaks in pipelines.	Analysis of the potential of glacial lake outburst floods.	Earthquake risk assessment.	Local air pollution assessment.
Stage II				
Social media	Promoting low-carbon lifestyles and information exchange.	Crowdsourcing data on vulnerable areas.	Pre- and post-disaster: sharing of information (e.g., finding missing persons).	Communication of sustainability strategies by corporations.
Apps	Calculation of a person's individual carbon dioxide (CO2) footprint and how to reduce it.	Enable rankings of household-level adaptation measures specific to the user's location.	Warning of extreme weather events in the vicinity of the user.	Recycling apps that show how and where to recycle a wide range of materials.
Cloud computing	Reduction needs for company's individual data center.	Computing power for big data analysis for adaptation planning.	Virtual mission continuity through independence from data centers that could be damaged by disasters.	Computing power for big data analysis (local air and water pollution data and their health effects).
Stage III				
Artificial intelligence, machine learning, and deep learning	Improving the prediction of energy demand and adjusting the capacity utilization of a power plant accordingly.	Better prediction of crop diseases which are becoming more prevalent with climate change.[c]	Improving the predictive analysis of the probability of natural hazards.	Improving resource efficiency in agriculture while improving crop yields.[d]
Internet of Things and smart systems	Optimization of vehicle traffic in cities.	Sensors to monitor soil quality in agriculture.	Sensor-based early-warning mechanisms (e.g., for earthquakes, floods).	Sensor-based detection of illegal logging and poaching.
Big data and predictive analytics	Collecting and analyzing data for household energy optimization.	Understanding the demand for food in a world with a growing population.	Evaluation of historical disaster data (e.g., of hurricanes) to identify patterns for future precise prediction models.	Identifying trends within different waste streams in order to take appropriate measures for reduction and better recycling.

continued on next page

Table 3 *continued*

Digital Technologies	Climate Change Mitigation	Climate Change Adaptation	Disaster Risk Management	Environmental Sustainability
Distributed ledger technology (DLT) and blockchain	Use of DLT for carbon markets (e.g., tracing and verification of emission reduction certificates).	Improving transparency in climate change adaptation measurements by regional authorities.	Avoiding inconsistencies and discrepancies when coordinating disaster relief efforts.	Use of DLT for sustainable supply chains.
Virtual and mixed reality	Low emission infrastructure planning. (e.g., improved representations of planned infrastructure adjustments lead to more precise resource utilization).	Visualization of climate change effects in urban areas to promote adaptation strategies.	Enable simulated disaster response training.[e]	Virtual representation of physical objects to save physical resources (e.g., product testing).
Robotics and unmanned vehicles	Drone-enabled production analytics for large solar farms. Heating cameras identify defect solar panels.[f]	Survey potential or effectiveness of adaptation measures in remote areas.[g]	Intervene where it becomes too dangerous for human helpers (e.g., collapsed structures).[h]	Autonomous plastic waste collection in rivers and oceans.[i]

[a] X. Lu et al. 2016. Unveiling hidden migration and mobility patterns in climate stressed regions: A longitudinal study of six million anonymous mobile phone users in Bangladesh. *Global Environmental Change*, Volume 38, 2016, Pages 1-7, ISSN 0959-3780, https://doi.org/10.1016/j.gloenvcha.2016.02.002. https://www.sciencedirect.com/science/article/pii/S0959378016300140.

[b] D.J. Wong, E. Jones and G.J. Rubin. 2017. Mobile text alerts are an effective way of communicatingemergency information to adolescents: Results from focusgroups with 12- to 18-year-olds. J Contingencies and Crisis Management.2018;26:183–192. DOI: 10.1111/1468-5973.12185. https://onlinelibrary.wiley.com/doi/pdf/10.1111/1468-5973.12185.

[c] PlantVillage is a free smartphone app used in Kenya and other African countries to diagnose diseases in crops such as cassava and potato. The app provides a diagnosis with a high level of accuracy talking with AI assistant, Nuru (Nuru means "light" in Swahili). In a test of machine learning (ML) models in the typical high light and temperature settings of an African farm, the app was found to perform twice better than human experts in making accurate diagnoses. Read more at https://bigdata.cgiar.org/inspire/inspire-challenge-2017/pest-and-disease-monitoring-by-using-artificial-intelligence/.

[d] Relevant technologies that are radically improving agricultural efficiency on the basis of AI, ML, and DL include blueradix.com and harvest-ai.com.

[e] Relevant current applications include, for example, by the US Department of Homeland Security (DHS) and the US Centers for Disease Control and Prevention (CDC). See also I. Lochhead and N. Hedley. 2017. Mixed reality emergency management: bringing virtual evacuation simulations into real-world built environments. International Journal of Digital Earth, Vol. 12, issue 2.

[f] Provided, for example, by raptormaps.com.

[g] This can be performed by drones specializing in Aerial Inspection Services for the agriculture sector. For example, Sense Fly (sensefly.com) has been experimenting with using drones for aerial inspection for agriculture, with obvious applications for monitoring agricultural adaptation measures.

[h] For example, the Tokyo Fire Department has been using remote-controlled robots (Robocue) for over a decade when rescuing disaster victims.

[i] For example, the Ocean Cleanup is a dutch start-up that works on solar-powered vehicles that autonomously collect plastic waste in rivers and oceans (theoceancleanup.com).

Examples of Digital Technologies in Applications in Climate Change Mitigation, Climate Change Adaptation, Disaster Risk Reduction Management, and Environmental Sustainability

Box 6: Climate Change Mitigation

Apps	Apps to reduce CO_2 emissions. Apps like Capture help their users to track their CO_2 emissions and help reduce them (photo by The Capture Club).
Project	**Reducing Carbon Dioxide (CO_2) Emissions with an App**
Challenge	The reduction of CO_2 emissions takes a big share in the efforts to mitigate climate change. A growing number of people want to contribute to this by making more sustainable life choices and creating lower emissions. Apps can represent fundamental help to do so.
Solution	The mobile App "Capture" helps its users to track their personal CO_2 footprint. This gives the user a sense of which activities cause particularly high or low emissions. The app automatically tracks emissions through means of transportation by global positioning system (GPS) tracking. Other emissions, for example, caused by food, have to be entered manually, but can be automated to some extent by an initial questionnaire. Based on recommendations of the Intergovernmental Panel on Climate Change (IPCC), the app communicates with its users about their personal CO_2 targets and associated progress. Furthermore, the app gives tips on how to reduce emissions and helps its users to offset their emissions by contributing to sustainable projects.[a]

[a]   Thecapture.club. 2020. Planet-friendly living, made possible. https://www.thecapture.club/.

Box 7: Climate Change Adaptation

Artificial Intelligence and Internet of Things	\n\n**Green Crops in Greenhouse.** Greenhouses can contribute to food security and adaptation to climate change (photo by Daniel Fazio).
Project	**Intelligent and autonomous greenhouses**
Challenge	Climate change threatens food security. Increasing temperatures, more extreme heat, and changed rain patterns are challenging farmers around the globe. Not only too little rain can be a consequence, but also too much, which may lead to floods.[a] Furthermore, a study showed that climate change is likely to contribute to crop loss from insects.[b]
Solution	One option is to invest in and develop better crops that are more resilient to these challenges. Another option is to increasingly rely on greenhouses for relevant crops. Greenhouses have the advantage that they can create favorable conditions and protect crops from external circumstances which cannot be influenced and are associated with climate change. Hydroponic systems even directly eliminate the use of soil, providing better use of resources through optimal nutrient solutions for improved resistance, growth, and pest control.\n\nModern greenhouses use Internet of Things (IoT) sensors that measure humidity, carbon dioxide (CO_2) content of the air, temperature, and the nutrition content in the water. In addition to this, camera systems can control the growth of crops. An innovation is to feed this data through an artificial intelligence (AI) algorithm which is now done by the start-up Harvest AI. The AI algorithm creates connections between optimally grown plants and humidity, CO_2 content of the air, temperature and nutrition content in the water, and adjusts these factors accordingly. Furthermore, it can predict at what time the crops can be harvested.[c]

[a]   Union of Concerned Scientists. 2019. Climate Change and Agriculture: A Perfect Storm in Farm Country. https://www.ucsusa.org/resources/climate-change-and-agriculture.

[b]   A. Wernick. 2018. Study: Climate change will bring more pests, crop losses. PRI. https://www.pri.org/stories/2018-09-23/study-climate-change-will-bring-more-pests-crop-losses.

[c]   Harvest AI. n.d. Next generation farming. https://harvest-ai.com/.

Box 8: Disaster Risk Management

Internet of Things

Flooding Street in Manila, Philippines. Internet of Things sensors can be used as early warning systems to give more response time to citizens and authorities (photo by Eric Sales).

Project	**Early warning systems against floods.** StormSense is a research project with the goal to develop a low-cost, energy-efficient, and scalable flood warning system.
Challenge	Flood warning systems may lack accuracy and are not yet implemented where they are needed. In addition, flood warnings may come with too little lead time or affected residents may miss relevant information.
Solution	More than 40 StormSense sensors have been installed at strategic points in a flood-endangered city in the United States. In addition, the system is able to ingest data from other sensors not owned by StormSense, but by government agencies. The sensors collect data for wind speed, air pressure, rainfall, and water levels and communicate with Internet of Things technology to report water levels. The project aims at improving flood depth predictions by at least 15% compared to competing systems. The data is processed and published in real-time and can be viewed online. In case of imminent flood risk, the system warns and informs authorities and also citizens directly as it can be connected to virtual assistants like Amazon Alexa. In addition, it can be used by city planners to build more resilient neighborhoods.[a]

[a] J.D. Loftis. 2016. StormSense. Virginia Institute of Marine Science. https://www.vims.edu/people/loftis_jd/StormSense/index.php.

Box 9: Environmental Sustainability

Unmanned Vehicles and Machine Learning	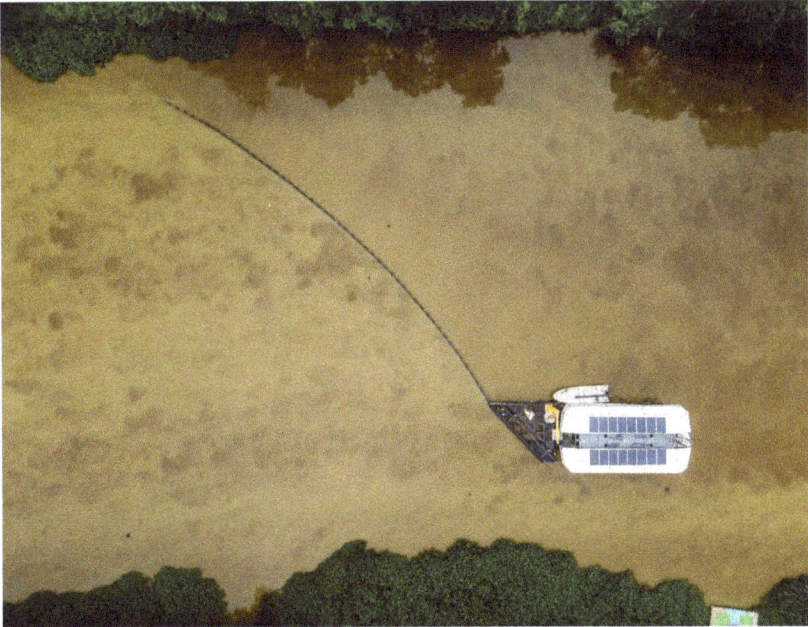**Unmanned vehicle in a river in Indonesia.** Unmanned vehicles, such as the Interceptor by the Ocean Cleanup, are used to collect waste from rivers and oceans (photo by The Ocean Cleanup).
Project	Unmanned vehicles that collect plastic waste from rivers.
Challenge	Big amounts of plastic pollute rivers and oceans and threaten marine ecosystems. "The Ocean Cleanup," a Dutch start-up plans to rid the oceans of plastic pollution. They realized that going after this waste requires large amounts of fossil fuels and, therefore, becomes expensive.
Solution	Hence, they decided to collect the plastic before it can enter the sea, in rivers and, therefore, developed the Inceptor, a solar-powered vehicle that autonomously collects garbage in rivers and prevents it from flowing into the sea. Currently, prototypes operate in rivers in Indonesia and Malaysia. However, not all objects floating in the water must be caught by the vehicle. Branches and leaves, for example, can flow to the sea without causing any harm. In addition, the vehicle's loading capacity would be exhausted too quickly if it were to collect too many such objects. The challenge is to teach the autonomous vehicle to identify what objects are harmful to the environment and what objects are not. Originally, the start-up had a staff member who labeled and identified pictures of plastic and other debris manually. This process was slow and inefficient. With the help of volunteers in a Microsoft hackathon, they developed a machine learning algorithm that teaches the program how to distinguish between plastic and other debris in the water on its own in a more efficient and faster way than a human could.[a]

[a] Business Insider. 2020. Microsoft teamed up with a nonprofit using autonomous 'interceptor' boats to clean up the ocean and is helping it identify trash with machine learning. https://www.businessinsider.com/microsoft-machine-learning-for-the-ocean-cleanup-project-2020-10.

In addition to the specific DTs presented in this publication, general trends, driven in part by digitization, can also be observed that tend toward a more climate-friendly society. Especially during the global COVID-19 pandemic, more workers were enabled to work remotely than before due to pandemic control measures. The management consulting firm McKinsey estimates that after the COVID-19 pandemic, almost a quarter of today's workforce could work remotely at least 3-5 days a week and be as efficient as working from the office. On the other hand, it expects that half of the workforce will not be able to work remotely because their work requires physical presence, such as agricultural or manufacturing work. Such occupations are more often found in emerging economies rather than developed economies. Hence, workers in advanced economies, such as Japan and the United States, can work remotely more often than workers in emerging economies, such as the PRC and India.[86]

Climate and environment can benefit from this trend, as fewer people have to commute to work every day. This reduces individual traffic, for example by car, and relieves public transport. Also, during the pandemic, a forced shift from business trips to videoconferencing and from teaching in educational institutions to e-learning happened. These shifts ultimately result in absolute reductions in emissions.

Detailed Analysis of Stage I Digital Technology

The Usefulness of Digital Technology from Stage I for Climate Change Adaptation, Climate Change Mitigation, Disaster Risk Management, and Environmental Sustainability

Stage I technologies are key in information provision on CCA, CCM, DRM, and environmental sustainability. This includes the actual capturing of information (through satellite imagery and GIS), its storing in online repositories and "searchability" in dedicated internet tools, and its communications in case of need via mobile phones, notably in the DRM context.

Technologies from Stage I in the Climate Mitigation Context

Among these, one might first consider information provision on sources of greenhouse gas (GHG) emissions and mitigation options. This includes online repositories and search engines on GHG emissions, mitigation options, or Data Support for Climate Change Assessments like the IPCC's TG-Data.[87] The internet's biggest asset is it provides access to information from various sources, enables real-time information, and can link content users and developers from different parts of the world—such that, for example, scientists can publish their latest findings on climate change impacts.

Second, considering the information on household-level choices for mitigation options, a range of products and services can be enabled by DT to provide mitigation options. These include tools to guide consumers among household energy savings options and household-level renewable energy applications like rooftop solar. Third, at the business level, a range of products and services can be enabled by Stage I-level DT to provide mitigation options. They also include searchable tools to guide business in their mitigation options, or searchable benchmarking websites on mitigation performance. Finally, information on national-level mitigation options include a range of the simple technologies outlined under Stage I included in design discussions for countries' nationally determined contributions (NDCs) under the Paris Agreement.[88]

---

[86]   McKinsey. 2020. What's next for remote work: An analysis of 2,000 tasks, 800 jobs, and nine countries. https://www.mckinsey.com/featured-insights/future-of-work/whats-next-for-remote-work-an-analysis-of-2000-tasks-800-jobs-and-nine-countries.

[87]   Intergovernmental Panel on Climate Change. IPCC Data. https://www.IPCC.ch/data.

[88]   World Bank. 2018. Blockchain and emerging digital technologies for enhancing post-2020 climate markets.

Technologies from Stage I in Adaptation Context

These include a range of the simple technologies outlined under Stage I. Information on climate change impacts and adaptation options includes online repositories for adaptation options, including online versions of National Adaptation Plans of Action. Information on household-level adaptation options include mobile phone alert systems in times of climate change-induced temporary water shortages. At the business level, DT-enabled mitigation options include, for example, in the agriculture sector, GIS-based frameworks for crop modeling and environmental matching procedures.[89]

A good example of using a previously existing database for effective adaptation goals is a UN database called Water Productivity which uses open access of remotely sensed derived data (Water Productivity through Open access of Remotely sensed derived data portal [WaPOR]).[90] It has been the main data source for a range of relevant apps, for example, PlantVillage Nuru, a free smartphone app used in Kenya and other African countries to diagnose diseases in crops such as cassava and potato. The app provides a diagnosis with a high level of accuracy talking with AI assistant, Nuru. More precisely, the WaPOR database uses the NASA's satellite-derived data and computes relevant metrics for crop productivity. Adaptation-relevant databases incorporated in PlantVillage Nuru include weather forecast data and a series of algorithms on adaptive measures that can be taken under certain conditions.[91]

Technologies from Stage I in Other Environmental Sustainability Areas

In line with the respective priorities outlined in OP3, this requires consideration of the Stage I DTs in air, water, and land pollution; biodiversity conservation; ecosystem management; and marine litter and/or blue economy.

Regarding air, water, and land pollution, relevant DTs include tools for air pollution information city-by-city for public policy decision-making (e.g., distribution of relevant public funds), but also DTs for individuals (e.g., to know when it is safe to go out or where to avoid going during days with severe levels of air pollution). Examples include remote sensing for airborne nitrogen dioxide monitoring over the PRC;[92] or air pollution information city-by-city in real-time, worldwide.[93] On the water pollution front, examples include reliable information, for example, for investment planning given water quality in different locations; or remote sensing to support water monitoring and management decisions.[94] With respect to land pollution, there are online repositories of toxic waste sites.

Regarding Stage I DTs for biodiversity conservation (wildlife trade, flora and fauna monitoring, and others), satellite-based work on tracking herds of endangered species is well established,[95] with more recent inventions (but still fairly basic Stage I-type technologies), including anti-poaching tags.[96]

---

[89] G. Fischer, M. Shah and H. van Velthuizen. 2002. Climate Change and Agricultural Vulnerability. IIASA, Laxenburg, Austria. http://pure.iiasa.ac.at/id/eprint/6670/.

[90] Food and Agriculture Organization (FAO). 2018. Water Productivity Through Open Access of Remotely Sensed Derived Data Portal (WaPOR). http://www.fao.org/3/CA1081EN/ca1081en.pdf.

[91] CGIAR. PlantVillage Nuru: AI for pest & disease monitoring. https://bigdata.cgiar.org/inspire/inspire-challenge-2017/pest-and-disease-monitoring-by-using-artificial-intelligence/ (accessed on 18 April 2021)

[92] NASA Earth Observatory. 2020. Airborne Nitrogen Dioxide Plummets Over China. https://earthobservatory.nasa.gov/images/146362/airborne-nitrogen-dioxide-plummets-over-china.

[93] Air Matters. n.d. https://air-quality.com/ (accessed 25 November 2020).

[94] Japitana et al. 2019. Catchment Characterization to Support Water Monitoring and Management Decisions Using Remote Sensing. Sustainable Environment Research (2019) 29:8. https://doi.org/10.1186/s42834-019-0008-5.

[95] G. Schofield. Satellite tracking large numbers of individuals to infer population level dispersal and core areas for the protection of an endangered species. Diversity and Distributions, vol 19 (7).

[96] P. O'Donogue et al. 2016. Real-time anti-poaching tags could help prevent imminent species extinctions. Journal of Applied Ecology, 53(1). pp. 5–10.

Moving on to ecosystem management (terrestrial ecosystems include forests and riverine and freshwater systems; marine ecosystems include corals, mangroves, seagrasses), informational Stage I DTs exist, for example, for tourism purposes on health of local forests or corals. While this has been done since the 1990s, new applications include experimentation with reef-scale thermal stress satellite products.[97]

In the marine litter and blue economy arena, much work has focused on satellite-based tracking of plastic pollution hot spots,[98] both for (so far, largely futile) ocean cleanup efforts, but also to identify key plastic waste source countries that need to reduce plastic leakage into the ocean.[99]

Technologies from Stage I in Disaster Risk Management

Currently, the use of DT in DRM is still strongly focused on information systems.[100] This stems partly from the fact that DRM information needs, on average, to be shared as widely as possible, and developers of DRM-relevant DTs, therefore, aim for the lowest common denominators of DT complexity. Many attempts to integrate DT into DRM in Asia have, therefore, been built so far on widespread access to mobile phones across the region, using weather monitoring-systems to deliver timely and near-universal warnings before disasters, and relief and recovery information after.

Two examples may help illustrate this. First, in the case of DRM risk assessment work in Jakarta, DRM agencies mapped the location of critical infrastructure using a tool (OpenStreetMaps) that allows volunteers to create digital maps that can be used without restrictions. Over a million buildings have been mapped in this way, with subsequent use of open-source software to analyze potential impacts of floods. This helped in the response to the 2013 and 2014 floods. Second, with regard to early warning, the Philippines' Project Nationwide Operational Assessment of Hazards (NOAH) uses real-time data from rain gauges, water sensors, and radar to provide evacuation alarms.[101]

Problems with Stage I Digital Technology

Among the various potential problems associated with Stage I DT, this publication focuses on two such problems. First, many Stage I DTs depend on the mining of raw materials such as cobalt or lithium, which have strong health and environmental side effects. For instance, in its Environmental, Social, and Governance data 2019, the world's largest cobalt-mining company, Glencore, reported that 17 people were killed in its operations in 2019.[102] Exposure to toxic pollution is reported to cause birth defects among the children of cobalt miners in the Democratic Republic of Congo, the world's biggest cobalt producing country.[103]

Second, electronic waste (e-waste) is the fastest growing waste stream in the world.[104] While electronics are made up of a sophisticated mix of valuable raw materials, their post-use extraction is inefficient and costly. Thus, through illegal waste trafficking between 25,000 tons (t)–37,000 t of e-waste mainly ends up in West

[97]   S. Heron et al. 2016. Validation of Reef-Scale Thermal Stress Satellite Products for Coral Bleaching Monitoring. *Remote Sensing 2016*. 8(1). p. 59.

[98]   Mongabay. 2020. *Satellite Imagery is Helping to Detect Plastic Pollution in the Ocean*. https://earth.org/satellite-imagery-helping-to-detect-plastic-pollution-in-the-ocean/.

[99]   J. Jambeck et al. 2015. Plastic waste inputs from land into the ocean. *Science*, Vol. 347, Issue 6223. 13 February 2015.

[100] Information technologies and disaster management – Benefits and Issues. https://www.sciencedirect.com/science/article/pii/S2590061719300122.

[101] World Bank. 2014. Third Flood Risk Management and Urban Resilience Workshop. 03–05 June. http://documents1.worldbank.org/curated/en/114171468245386234/pdf/906430WP0P13010roceedings0June02014.pdf.

[102] Glencore. 2020. Environmental, Social and Governance data 2019. https://www.glencore.com/dam/jcr:54944abf-082a-4d5b-80e7-66501a5db876/2019-ESG-Databook-.pdf.

[103] Amnesty International. 2020. DRC: Alarming research shows long lasting harm from cobalt mine abuses. 06 May. https://www.amnesty.org/en/latest/news/2020/05/drc-alarming-research-harm-from-cobalt-mine-abuses/.

[104] R. Leblanc. 2019. E-Waste and the Importance of Electronics Recycling. *The Balance Small Business*. 25 June. https://www.thebalancesmb.com/e-waste-and-the-importance-of-electronics-recycling-2877783.

Africa (Ghana and Nigeria) and Asia (Bangladesh; India; Hong Kong, China; Pakistan; the PRC; and Viet Nam), where the waste is recycled under conditions that are harmful to health and the environment.

Detailed Analysis of Stage II Digital Technology

The Usefulness of Digital Technology from Stage II for Climate Change Adaptation, Climate Change Mitigation, Disaster Risk Management, and Environmental Sustainability

The use of Stage II DT for the issue areas under discussion in this publication are mostly from these enabling interaction through dedicated types of social media—i.e., collaboratively created climate change or sustainability solutions on social media. Another such way is in by using apps, for example, in response to disasters or guidance to reduce one's individual CO_2 footprint.

Technologies from Stage II in the Climate Mitigation Context

This includes apps and social media that promote sustainable consumption, including apps that calculate the carbon footprint and other personal environmental-climate impacts. "Coordination apps" and social media for sustainability services encompass platforms on social media channels for education about climate change. The landscape of these is too vast to discuss here, but includes various apps for reducing a person's carbon footprint, for example, through different forms of personal adjustments (modes of travel, food consumption). Arguably one of the most sophisticated sets of complex interactive tools is climateinteractive.org/. Grown out of the Massachusetts Institute of Technology and based on system dynamics modeling, it allows users to conduct simulations, and play out complex mitigation scenarios.

Technologies from Stage II in the Adaptation Context

Stage II-type technologies may support adaptation planning through app-based information exchange. In the agriculture sector, it may mean discussion fora for farmers on experiences with new crop varieties under hotter and/or dryer climatic conditions. For instance, in Qixing farm in Northeast PRC's Heilongjiang province, a modern big data center has transformed the Qixing farm, which is the largest paddy farm in the province covering 81,300 hectares. The center uses data collected from high-resolution Gaofen-1 satellites as well as data obtained from meteorological machines and those related to land and the environment. IoT data are obtained from meteorological machines, underground water-level monitoring equipment, and other sources.[105] Farming decisions are made based on real-time data related to temperature, humidity, wind direction, soil temperature, and humidity. Farmers need to install a mobile app to access the real-time data. The app, "Modern Agriculture Platform (MAP)" was developed by state-owned enterprise Sinochem Group.[106]

Technologies from Stage II in Disaster Risk Management

Many of the Stage II DTs used in DRM consist of disaster-focused apps, and social network-based information and/or alert systems. This may include apps and social networks allowing for citizens to inform authorities about weak points in civil protection (e.g., flood protection); or to be informed about natural hazards (early warning, shelter options, relief coordination).

---

[105] Xinhua. 2018. *Xinhua Headlines: Big data reshaping harvest for Chinese farmers.* 29 November. http://www.xinhuanet.com/english/2018-11/29/c_137640065_2.htm.

[106] CGTN. 2019. *Feeding 1.4 Billion: Smart farming in China's big grain warehouse.* 13 December. https://news.cgtn.com/news/2019-12-13/Feeding-1-4-Billion-Smart-farming-in-China-s-big-grain-warehouse-MohBFcaajK/index.html.

There is also massive use of generic Stage II-DTs in the DRM field, as evidenced by the high rate of Twitter usage of crowd-map, for example, flood locations in real-time during the monsoon season in many Asian countries.

While not social media per se, open-source software is, by definition, collaborative and has been used frequently as a cheap alternative to conventional software—allowing high-level customization encourages cross-fertilization, community building, and code sharing with local developers. This has led, for example, to collaboration with experts in Haiti or Indonesia on DRM-related DT tools.[107]

Technologies from Stage II in Other Environmental Sustainability Areas

These are the sustainability focal areas from OP3, starting with air, water, and land pollution.[108] On the air pollution front are apps to track air pollution and possible effects on vulnerable groups, such as IQAir–the World Air Quality Recommendation for Sensitive Groups.[109] Similarly, there are apps to track water pollution and its possible effects on vulnerable groups. An example is the blue map, an air and water pollution source map.[110] In the land pollution arena, it means apps to allow waste collection and disposal: for instance, smartphone apps to manage household waste in Indonesia.[111]

As for biodiversity conservation (wildlife trade, flora and fauna monitoring, and others), a key category of Stage II technologies includes coordination apps for anti-poaching efforts. For instance, Thailand introduced the Spatial Monitoring and Reporting Tool (SMART) tech to protect Asian elephants: SMART allows its users to identify hot spots where attention is needed. Combined with the app, Cybertracker, users can immediately and thoroughly collect data, which can later be uploaded and evaluated with SMART.[112]

In ecosystem management (terrestrial ecosystems include forests and riverine and freshwater systems; marine ecosystems include corals, mangroves, seagrasses), examples of Stage II DT include social media for ecosystem management institutions. For instance, coral reef management apps allow divers to share photos and data about what they encounter when visiting a reef which, in turn, data scientists and policy makers can use when conducting coral ecosystem management.[113]

An example in the context of marine litter and blue economy comes from Indonesia, one of the world's most important source countries of plastic waste given its large combined coastline of over 13,000 islands. There are household waste apps meant in part to help manage household plastic waste and its uncontrolled flow into the ocean (footnote 111).

Problems with Scaling Stage II Digital Technology

As with Stage I DTs, problems abound with the increased use of Stage II DTs. First, these DTs almost invariably infringe on data protection preferences and its users' privacy. For instance, according to a poll of 217 information

---

[107]   World Bank. 2016. *World Development Report 2016*: Digital Dividends. Washington, DC: World Bank.

[108]   The categories mentioned are taken from the Life Cycle Assessment impact categories. According to the Standard DIN EN ISO 14040, impact categories are classes representing environmental issues of concern to which the results of life cycle inventory analysis may be assigned. European Commission-Joint Research Centre and Institute for Environment and Sustainability. 2011. *International Reference Life Cycle Data System (ILCD) Handbook- Recommendations for Life Cycle Impact Assessment in the European context*. Luxemburg: Publications Office of the European Union; Stockholm University. 2020. *The nine planetary boundaries*. https://www.stockholmresilience.org/research/ planetary-boundaries/planetary-boundaries/about-the-research/the-nine-planetary-boundaries.html.

[109]   IQAir. n.d. Air Visual App. https://www.iqair.com/us/air-quality-app.

[110]   IPE. n.d. Blue Map. http://wwwen.ipe.org.cn/appdownload30_en/pc/index.html.

[111]   A. Pamungkas. 2019. *Indonesia: A smartphone app to manage household waste*. Deutsche Welle. https://www.dw.com/en/indonesia-a-smartphone-app-to-manage-household-waste/av-51558686.

[112]   IUCN. 2019. Thailand introduces SMART tech to protect Asian elephants. https://www.iucn.org/news/asia/201909/thailand-introduces-smart-tech-protect-asian-elephants.

[113]   This system was pioneered for Australia's Great Barrier Reef (http://www.gbrmpa.gov.au/our-work/eye-on-the-reef), but has since been experimented with in various Asian coral reef systems as well.

technology (IT) professionals conducted by network visibility and traffic monitoring firm, Gigamon, at the Infosecurity Europe 2019 conference, most respondents believed that social media applications were responsible for bringing in the most malware in their enterprises.[114] Likewise, as noted in the 2019 Verizon Data Breach Investigations Report (DBIR), 33% of external attacks used social media to launch cyberattacks.[115] About a third of those attacks involved malware.[116]

Second, running DTs requires large amounts of energy for the data centers. For instance, in 2018, the data centers worldwide were estimated to consume 205 terawatt-hours of electricity, which was around 1% of the world's total electricity consumption.[117]

Third, the number of app or social media platform providers of each app or social media platform is too complex for proper quality control. Fourth, for any given app, quality control by individual users is equally difficult. Fifth, there is a need for considerable initial minimal network size for any useful apps or social media platforms related to climate change.

Detailed Analysis of Stage III Digital Technology

The Usefulness of Digital Technology from Stage II for Climate Change Adaptation, Climate Change Mitigation, Disaster Risk Management, and Environmental Sustainability

These include, for instance, a move by various public and private entities toward autonomous DT systems AI, where it is not humans who interpret the data obtained from the internet, but the system interprets these, responds independently, and (through ML) learns over time and thereby creates self-improving systems, with obvious efficiency gains at the household or firm-level that have, for example, GHG emissions mitigation benefits. There are also verification systems for environmental data streams through distributed ledger technologies (DLT) or traditional databases. Arguably, at the most systemic level is complex interconnected DT systems, notably the Internet of Things (IoT), which has a range of applications for creating, for example, energy- or water-efficient systems at the household or business level, as earlier discussed about the IoT.

Technologies from Stage III in the Climate Mitigation Context

Various Stage III DTs have the strong potential to support climate mitigation efforts. For instance, the application of AI levers could reduce worldwide GHG emissions by 4% in 2030.[118] As another example, DLT can provide more transparency about individual and collective climate action, as well as reliable provenance of climate-friendly products within global supply chains. They can also incentivize climate-friendly behavior with purpose-driven tokens. IoT can enable the establishment of connected sensors that can deliver data from various fields and for various emissions-intensive appliances, such as smart home energy management system (EMS) using IoT and big data analytics approach to promote energy efficiency.[119]

---

[114] W. Ashford. 2019. Social media and enterprise apps pose big security risks. ComputerWeekly.com. https://www.computerweekly.com/news/252469873/Social-media-and-enterprise-apps-pose-big-security-risks.

[115] Verizon. 2019. 2019 Data Breach Investigations Report. https://enterprise.verizon.com/resources/reports/dbir/.

[116] T. Bradley. 2020. Cybersecurity Priorities are A Matter Of Perspective. Forbes. https://www.forbes.com/sites/tonybradley/2020/02/05/cybersecurity-priorities-are-a-matter-of-perspective/#5575d6345d17.

[117] Informa Data Center Knowledge. 2020. Study: Data Centers Responsible for 1 Percent of All Electricity Consumed Worldwide. https://www.datacenterknowledge.com/energy/study-data-centers-responsible-1-percent-all-electricity-consumed-worldwide.

[118] PwC. 2020. *How AI can enable a sustainable future*. London. https://www.pwc.co.uk/services/sustainability-climate-change/insights/how-ai-future-can-enable-sustainable-future.html.

[119] Example on this from Arab Gulf countries. A.R. Al-Ali et al. 2017. A smart home energy management system using IoT and big data analytics approach. *IEEE Transactions on Consumer Electronics*. November. Vol. 63 (4). pp. 426–434.

Technologies from Stage III in the Adaptation Context

Stage III DT can support climate adaptation planning and processes in various ways. For instance, IoT can aid water level predictions via sensors in coastal areas—as the StormSense research project does to predict floods. This system can even be connected to virtual home-assistant speakers to warn the inhabitants of a flood.[120] To give another example, the use of artificial intelligence for agricultural systems affected by diminished water availability, thus "learning" maximum efficient water use in a specific location. One such solution was developed by the research organization, DHI GRAS. It combines satellite data (that measures how much water evaporates from agricultural land) with meteorological data, and feeds this information to an AI algorithm. Farmers are then advised by the system on the correct amount of water needed to efficiently water the crops.[121]

Technologies from Stage III in Disaster Risk Management

In DRM as well, the potential of Stage III technologies (Box 10) is significant. It includes the use of AI and ML in analyzing potential hazards, robotics in the recovery of victims from collapsed structures and coordination of relief efforts in post-disaster zones,[122] or use of remote sensing for hazard analysis. For instance, in the case of the 2015 Nepal earthquake, the US Geological Survey (USGS) responded to the crisis by providing landslide-hazard expertise to Nepalese agencies and affected villages. In addition to collaborating with an international group of remote-sensing scientists to document the spatial distribution of landslides in the first few weeks following the earthquake, the USGS conducted in-country landslide hazard assessments.[123] Another example of Stage III technologies in DRM is that connected sensors and/or IoT may help in providing water level and flood warnings as well as foresee other disasters such as earthquakes and potential landslides in prone areas, assisting the civilians and authorities to take drastic action on such issues.[124]

Another field of application consists of damage assessment and response via unmanned vehicles, thus alleviating the well-documented difficulties of matching relief supplies with need in the chaos following a disaster. For instance, after the 2015 Vanuatu cyclone, drones were used to survey affected areas, which provided more detailed damage information than was possible with an aerial photograph.[125]

Technologies from Stage III in Other Environmental Sustainability Arenas

On the air pollution front, a key Stage III-type invention is smart sensing for efficient traffic flow and in industrial or agricultural air pollution issues. This includes, for example, Breeze Technologies' Air Quality Sensors, with air quality data uploaded in the cloud for further processing;[126] and Smart Traffic Management to regulate city traffic, including sensors and traffic signals for monitoring, controlling, and reacting to traffic conditions.[127]

---

[120]   J.D. Loftis. 2016. *StormSense*. Virginia Institute of Marine Science. https://www.vims.edu/people/loftis_jd/StormSense/index.php.

[121]   Microsoft News Centre Europe. n.d. How AI and satellite data are helping farmers waste less water. https://news.microsoft.com/europe/features/how-ai-and-satellite-data-are-helping-farmers-waste-less-water/.

[122]   M. Reynaert. n.d. *Drones: Propelling Sustainable Development*. RESET. https://reset.org/node/27732.

[123]   US Geological Survey. 2015. Assessment of existing and potential landslide hazards resulting from the 25 April 2015 Gorkha, Nepal earthquake sequence.

[124]   BBVA. 2019. *The Internet of things and its impact on sustainability*. 21 November. https://www.bbva.com/en/the-internet-of-things-and-its-impact-on-sustainability/; EcoMENA. 2020. *The Role of IoT in Sustainable Development*. https://www.ecomena.org/internet-of-things/; B. Buntz. 2019. How IoT Technology Can Help the Environment. *IoTWorldToday*. 19 December. https://www.iotworldtoday.com/2019/12/19/how-iot-technology-can-help-the-environment/.

[125]   B.C. Howard. 2015. Vanuatu Puts Drones in the Sky to See Cyclone Damage. *National Geographic*. https://www.nationalgeographic.com/news/2015/04/150406-vanuatu-cyclone-pam-relief-drones-uavs-crisis-mapping-patrick-meier/.

[126]   Breeze Technologies. https://www.breeze-technologies.de/solutions/urban-air-quality/.

[127]   HEREmobility. *Smart Traffic Systems 101: Components, Benefits, And The Big Data Connection*. https://mobility.here.com/learn/smart-transportation/smart-traffic-systems-101-components-benefits-and-big-data-connection#pgid-1656.

The water pollution field has notably seen great progress in smart sensing solutions, such as with smart sensor networks for seawater quality monitoring[128] or improving the health of marine ecosystems by automatically detecting potential oil leaks.[129] In land pollution, notable progress for the Stage III DTs includes the combination of robotics with AI to sort hazardous waste.[130]

Box 10: Using Stage III Digital Technology in the Recovery from COVID-19 in the People's Republic of China

After an environmental disaster, it is expected that economic activities start recovering. However, different economic activities may recover at different paces, which can be captured by combining artificial intelligence, digital technology (DL), and remote sensing tools. By analyzing satellite images of activities such as movements of animals and how **agricultural lands** are being used, it is possible to know if economic activities are recovering from the distress. The resulting insights can be used to take policy measures that can be better targeted to address the most serious problems.

For example, online banking company WeBank of the People's Republic of China (PRC) used a neural network to analyze images from various satellites, including the Sentinel-2 satellite. The images made it possible for the system to observe steel manufacturing inside a plant. A deep learning framework known as SolarNet, which was originally designed to detect solar farms using large-scale satellite imagery data was used for this purpose.[a] The analysis indicated that in the early days of the coronavirus disease (COVID-19) outbreak, steel manufacturing had dropped to 29% of capacity. By 9 February, it recovered to 76%.[b] To build a complete picture of the state of manufacturing and commercial activity, the system combined satellite imagery with global positioning system data from mobile phones and social media posts. Using the data gathered by the system, it predicted that most Chinese workers outside Wuhan would return to work by the end of March. By counting cars in large corporate parking lots in Shanghai, it found that Tesla's Shanghai car production had fully recovered by 10 February, but Shanghai Disneyland and other tourist attractions were shut down (footnote b).

WeBank analyzed visible, infrared, near-infrared, and short-wave infrared images from Sentinel-2 and other satellites.[d] These are mainly in the 10 meter to 30 meter resolution range, which are considered to be medium-resolution imageries.[e]

[a] X. Hou et al. 2019. SolarNet: A Deep Learning Framework to Map Solar Power Plants in China From Satellite Imagery. https://arxiv.org/abs/1912.03685.
[b] T. S. Perry. 2020. Satellites and AI Monitor Chinese Economy's Reaction to Coronavirus. https://spectrum.ieee.org/view-from-the-valley/artificial-intelligence/machine-learning/satellites-and-ai-monitor-chinese-economys-reaction-to-coronavirus.
[c] A. Johnson. 2020. How Artificial Intelligence is Aiding the Fight Against Coronavirus. https://www.datainnovation.org/2020/03/how-artificial-intelligence-is-aiding-the-fight-against-coronavirus/.
[d] GIS Geography. 2020. *Sentinel 2 Bands and Combinations*. https://gisgeography.com/sentinel-2-bands-combinations/.
[e] Earth Observing System. 2019. *Satellite Data: What Spatial Resolution Is Enough?* https://eos.com/blog/satellite-data-what-spatial-resolution-is-enough-for-you/.

---

[128] F. Adamo et al. 2015. A Smart Sensor Network for Sea Water Quality Monitoring. *IEEE Sensors Journal* 15(5). pp. 2,514–2,522.
[129] R. Vinuesa et al. 2020. The role of artificial intelligence in achieving the Sustainable Development Goals. *Nat Commun* 11, 233. https://doi.org/10.1038/s41467-019-14108-y.
[130] Danish Technological Institute. n.d. *Robots with Artificial Intelligence to Sort Hazardous Waste*. https://www.dti.dk/specialists/robots-with-artificial-intelligence-to-sort-hazardous-waste/38310.

In the biodiversity conservation arena, progress is seen with sensors on endangered species for tracking migratory patterns and poaching.[131] An example includes Forest Guard, a Smart Sensor System that detects and prevents illegal logging and wildfires by recognizing saw sounds and fireplace sounds.[132] If the system recognizes such sounds, it sends an emergency signal including the location to the respective authorities.[133]

With respect to ecosystem management (terrestrial ecosystems include forests and riverine and freshwater systems; marine ecosystems include corals, mangroves, seagrasses), robotics has increasingly been used for monitoring purposes. For instance, the introduction of robotics into forestry management is now easily a decade old,[134] and recent advances have also been made, for example, in applying AI to forestry monitoring.[135]

Finally, regarding marine litter and blue economy, key progress with respect to Stage III DT is represented by robotics-based environmental monitoring of garbage patches and attempts at subsequent collection.[136] Solar-powered unmanned vessels can also collect data to monitor fish stocks, collect climate and weather data, and replace fuel-powered, non-autonomous boats.[137]

Problems with Scaling Stage III Digital Technology

A number of problems can potentially impede the scaling of Stage III DTs. Their use for climate change would require a life cycle assessment of their net carbon impact.

For IoT, for instance, a fragmented landscape for standardization prevents interoperability. A second concern is that some embedded devices are expensive. Most IoT devices are also prone to cyberattacks and privacy violations because not enough consideration is given to security in their design. For instance, it is not practical to install a firewall on inexpensive devices due to insufficient memory.

A key challenge that arises in the implementation of DLTs and blockchain consists of bringing all the relevant parties together, which can be a difficult undertaking in many cases (footnote 78). The network effects cannot be captured without the participation of all relevant parties. Many proposed uses of DLT can also be effectively delivered via traditional databases, too, which may be simpler and easier to implement.

Stage III technologies usually imply considerable up-front costs and a large amount of computing power and hence energy consumption.[138] These challenges are especially severe for ADB's developing member countries (DMCs).

The lack of capabilities such as technical knowledge and skills also affect the deployment of Stage III technologies (footnote 78). For instance, as of 2018, there were about 20 million software developers in the

[131] Combined with technologies such as Radio Frequency Identification (RFID), the technology behind contactless payment with bank cards, IoT can be used to prevent poaching of rhinoceroses. RFID chips are attached to the animals' horns and enables drones to easy identify the animal on the ground and track their movements and migration patterns. If the horn is detached from the rhino, it helps authorities to catch poachers which might create an deterrent effect on other poachers. Fast Company. 2014. *RFID-Tagged Rhinos and Smart Watering Holes: The Google-Funded Tech Fighting Poaching.* 02 July. https://www.fastcompany.com/3026125/rfid-tagged-rhinos-and-smart-watering-holes-the-google-funded-tech-fighting-poaching; I. Lacmanović. B. Radulović and D. Lacmanović. 2010. *Contactless payment systems based on RFID technology.* Conference Paper for the 33rd International Convention MIPRO. Croatia. 24–28 May.

[132] Forest Guard. https://www.ggf.lu/press/news/news-detail/clim-competition-2020-announces-15-finalists-with-circular-economy-solutions-the-most-popular-theme-30.

[133] Forest Guard. https://forestguard228852108.wordpress.com/.

[134] K. Bayne et al. 2012. The introduction of robotics for New Zealand forestry operations: Forest sector employee perceptions and implications. *Technology in Society.* Vol. 34 (2).

[135] J. Sandino et al. 2018. Aerial Mapping of Forests Affected by Pathogens Using UAVs, Hyperspectral Sensors, and Artificial Intelligence. *Sensors 2018.* 18(4). p. 944.

[136] Clear Blue Sea. n.d. *Foating Robot for Eliminating Debris.* https://www.clearbluesea.org/meet-fred/.

[137] Greentown Labs. n.d. *Datamaran: A Satellite for the Seas.* Autonomous Marine Systems. https://greentownlabs.com/members/autonomous-marine-systems/.

[138] S. Ejiaku. 2014. Technology Adoption: Issues and Challenges in Information Technology Adoption in Emerging Economies. https://core.ac.uk/download/pdf/55335431.pdf.

world, but only 0.1% of them knew about blockchain (footnote 78). No more than 6,000 coders were estimated to have the levels of skill and experience needed to develop high-quality blockchain solutions.[139] Out of India's 2 million software developers, only 5,000 were estimated to have blockchain skills.[140]

Similar challenges have been noted in AI. For instance, India is estimated to have about 50 to 75 AI researchers.[141] According to the employability evaluation and certification company Aspiring Minds' Annual Employability Survey 2019, only 2.5% of Indian engineers had AI skills required by the industry.[142] India has also faced a severe shortage of qualified faculty members to teach AI courses in its universities.[143]

Combining Different Digital Technologies to Achieve Amplified Impacts

A DT is likely to have amplified benefits beyond its direct effects in terms of achieving various goals related to CCM, CCA, environmental sustainability, and DRM if it is combined with other DTs. For instance, Box 11 shows the benefits of utilizing a combination of DTs to support decarbonization in the transport and power generation sectors.

Box 11: Shifting to Low-Carbon Transport and Power Generation

Another shift that has been in evidence for some time now, is away from internal combustion vehicles toward electric vehicles (EV). The management consulting firm Deloitte expects that annual global EV sales will go up to 31.1 million by 2030 compared to 2.5 million in 2020.[a] To ensure sustainable benefits from this shift, electricity required to run EVs must mostly come from renewable sources in the future and be distributed by an efficient grid system/smart grid. Among other benefits to traditional grids, they imply real-time communication between the participants in the network, for example, from the energy demand side to the energy production side. Hence, they can react quickly to varying energy consumption and thus reduce energy waste.[b] In this way, smart grids can efficiently distribute the varying feeds of energy from renewable sources like wind and solar energy.[c] Smart grids require several digital technologies to work. The data from Internet of Things sensors in combination with big data analysis and artificial intelligence not only help to detect inefficiencies and faulty devices in the system, but also improve weather forecasts for estimating the potential feed-in of renewable energies.[d]

ᵃ   Deloitte. 2020. *Electric Vehicles – Setting a course for 2030*. Electric vehicle trends. Deloitte Insights. https://www2.deloitte.com/us/en/insights/focus/future-of-mobility/electric-vehicle-trends-2030.html.
ᵇ   Digiteum. 2019. The Role of IoT in Smart Grid Technology. https://www.digiteum.com/iot-smart-grid-technology.
ᶜ   IEEE. 2019. *The Smart Grid and Renewable Energy*. IEEE Innovation at Work. https://innovationatwork.ieee.org/smart-grid-transforming-renewable-energy/.
ᵈ   Ye. L. and Zhao. Y.N. n.d. *Wind power prediction in the Smart Grid*. https://www.springeropen.com/p/engineering/smart-grid/wind-power-prediction-in-the-smart-grid.

High levels of accuracy have been achieved using satellite imagery data combining with ML. The global security and aerospace company, Lockheed Martin, has claimed that its system has an accuracy of over 90% to identify characteristics of an object area or target. This accuracy is comparable to the commonly used deep learning algorithm in the medical field known as convolution neural network (CNN) to assist disease diagnosis, which is

---

[139]   P. Suprunov. 2018. How much does it cost to hire a blockchain developer? *Medium*. https://medium.com/practical-blockchain/how-much-does-it-cost-to-hire-a-blockchain-developer-16b4ffb372e5.

[140]   M. Agarwal. 2018. Blockchain: India likely to see brain drain as 80% developers may move abroad. Inc42. https://inc42.com/buzz/blockchain-india-likely-to-suffer-brain-drain-as-80-developers-prepare-to-move-abroad/.

[141]   R. Gupta. 2019. The State Of Artificial Intelligence Development in India. *ViaNews*. https://via.news/asia/artificial-intelligence-development-india/.

[142]   Business Today India. 2019. *80% of Indian engineers not fit for jobs, says survey*. 25 March. https://www.businesstoday.in/current/corporate/indian-engineers-tech-jobs-survey-80-per-cent-of-indian-engineers-not-fit-for-jobs-says-survey/story/330869.html.

[143]   S. Ravi and P. Nagaraj. 2018. Harnessing the future of AI in India. Brookings Institute. https://www.brookings.edu/research/harnessing-the-future-of-ai-in-india/.

reported to yield an accuracy level of over 90% in diagnosing and providing treatment suggestions.[144] Lockheed Martin's self-learning model recognizes ships, airplanes, buildings, seaports, and many other commercial categories.[145] Likewise, in a study that combined satellite imagery and deep learning to identify communities in rural and remote settings, the researchers reported that they achieved positive predictive value of 86.47%.[146] Identifying communities and various infrastructures in rural and remote settings is extremely helpful in DRM planning.

People affected by and responding to a disaster need to know the locations of hospitals, pharmacies, and stores. It is also important to see transport infrastructures such as roads and railways to make food and medicine available to the needy population. They must be able to count the houses to know the population so that aid workers can decide the number of vaccines to bring.[147] During COVID-19, the Humanitarian OpenStreetMap Team (HOT) volunteers' tasks included finding hospitals in countries such as Iran, the Philippines, and Turkey, including relevant information about business and practice details such as addresses, opening hours, services provided, outlines of hospital buildings, and helipads in places that use air ambulances.[148]

Some companies are developing solutions involving satellite images, blockchain, and AI that reward sustainable farming practices. One such example is Oracle's partnership with the World Bee Project to help farmers manage the bee population and pollinator habitats (footnote 78). The plan is to take images of the farm with drones or satellites and utilize AI-based image recognition algorithms to evaluate whether the way farmland is managed supports bee colonies and other pollinators in a sustainable way.[149] Research has indicated that farms that allocate a certain proportion of their land to plant flowering crops such as spices, oilseeds, buckwheat, and sunflowers can increase crop yields by up to 79% due to efficient pollination from bees (footnote 78). Another idea that is being explored in this project is to break down the DNA sample of the honey and store it in a blockchain, allowing the producers to certify that the honey hasn't been tampered across the supply chain (footnote 149).

Indirect Causal Chains

Various indirect causal chains associated with the use of DTs can also contribute to CCM, CCA, DRM, and environmental sustainability. DTs can be used to bring changes in policy, market, and industry dynamics. For instance, data generated by DTs can be effectively used as evidence to pressure policy makers to address environmental issues that they would prefer to ignore, for political or other reasons. This publication illustrates this point below by considering how environmental activists have used satellite data to bring about policy changes.

Every day, earth observation satellites gather huge amounts of data about the earth's physical, chemical, and biological characteristics also known as the Earth system. The data gathered by satellites are used for civil, military, and commercial activities. These satellites provide the same level of richness of data for remote areas of least developed countries as well as urban areas of developed countries. Archival satellite data can help establish a cause-and-effect linkage between events, which can help determine the causes of environmental degradation and problems and identify environment offenders. Activist groups such as Greenpeace, Environmental

[144]   F. Jiang et al. 2017. Artificial intelligence in healthcare: past, present and future. *Stroke and Vascular Neurology*. https://svn.bmj.com/content/2/4/230.full.

[145]   I. Singh. 2019. Lockheed Martin develops AI model for satellite imagery analysis. 18 June.https://geoawesomeness.com/lockheed-martin-artificial-intelligence-model-satellite-imagery-analysis/.

[146]   E. Bruzelius et al. 2019. Satellite images and machine learning can identify remote communities to facilitate access to health services. *Journal of the American Medical Informatics Association*. 26 (8–9). pp. 806–12. https://doi.org/10.1093/jamia/ocz111.

[147]   S. Scoles. 2020.Satellite Data Reveals the Pandemic's Effects From Above. *Wired*. 9 April. https://www.wired.com/story/satellite-data-reveals-the-pandemics-effects-from-above/.

[148]   HOT. 2020. Volunteer Mappers. https://www.hotosm.org/volunteer-opportunities/volunteer-mappers/.

[149]   J. Charness. 2019. How Oracle and The World Bee Project are Using AI to Save Bees. Oracle AI and Data Science Blog. https://blogs.oracle.com/datascience/how-oracle-and-the-world-bee-project-are-using-ai-to-save-bees-v2.

Investigation Agency (EIA), and Conservación Amazónica have used archival satellite data to monitor changes in the land surface and create sequence of events such as deforestation, land-clearing wildfire, and land development.

Consider the 2015 Indonesian fires, which represent one of the "worst manmade environmental disaster since the BP gulf oil spill."[150] Thousands of fires were deliberately started to clear land for palm oil and paper products. During January–October of 2015, over 117,000 forest fires had been detected via satellite in Indonesia, most of which were suspected to be started deliberately to clear land for farming. It was estimated that in 2015, Indonesian fires had emitted roughly 1,713 million metric tonnes of equivalents of carbon dioxide as of 9 November. These estimates come from data collected by satellites, which can sense fires. Some examples include Germany's TET-1 and NASA's MODIS, which covers the whole world and transmits data 24/7. Active fires emit radiations that can be picked up by the satellites on bands that are dedicated to this purpose. From this information, scientists can calculate the size, temperature, and number of fires.

Indonesian and foreign media used the findings from the World Resources Institute (WRI), the Global Fire Emissions Database (GFED), and other resources to advocate for the Indonesian government to be more responsive to the tragedy, act on the issue, and adopt better peatland management practices.[151] Environment activist group, Greenpeace, collected and presented evidence in a clear and thorough fashion. It also released video footage taken from drones. Greenpeace Indonesia drones found that land burned in the autumn of 2015 on the island of Borneo were turned into palm oil plantations after a few weeks.[152] Greenpeace researchers examined about 112,000 fire hot spots recorded from 1 August to 26 October 2015, which showed that about 40% of the fires had taken place inside so-called mapped concessions, which is land granted by the government to companies for logging or plantation development. Greenpeace researchers also found that Asia Pulp & Paper, the largest concession holder in Indonesia, was the company associated with most of the fires.

DT-based evidence is likely to be more convincing and to eventually prompt policy changes. Such DT-based activisms have helped produce some desirable outcomes. In October 2015, Indonesian President Joko Widodo instructed the environment and forestry minister to stop issuing new permits on peatlands and immediately begin revitalization.[153] Singapore issued legal notices to Asia Pulp & Paper and four other Indonesian companies whose concessions are full of fires causing air pollution across the region.[154] According to Singapore's Transboundary Haze Pollution Act of 2014, foreign companies can be held responsible for polluting air and can be fined up to $1.4 million (footnote 154).

---

[150]   New Europe. 2015. Indonesia's devastating forest fires are manmade. 9 November. http://neurope.eu/article/indonesias-devastating-forest-fires-are-manmade/ (accessed 17 February 2016).

[151]   N. Kshetri. 2016. *Big Data's Big Potential in Developing Economies: Impact on Agriculture, Health and Environmental Security*. Wallingford, Oxon, UK: Centre for Agriculture and Biosciences International (CABI) Publishing.

[152]   N. Coca. 2015. Palm Oil from Freshly-Burned Land: Coming to a Grocery Store Near You. 13 November. http://www.triplepundit.com/2015/11/palm-oil-freshly-burned-land-coming-grocery-store-near/ (accessed 11 October 2020).

[153]   *Mongabay*. 2015. Jokowi pledges Indonesia peatland 'revitalization' to stop the burning. 30 October. http://news.mongabay.com/2015/10/jokowi-pledges-greater-indonesia-peatland-revitalization-no-legal-breakthrough-yet/ (accessed 11 October 2020).

[154]   *Mongabay*. 2015. Singapore takes legal action against 5 Indonesian companies over haze. 1 October. http://news.mongabay.com/2015/10/singapore-takes-legal-action-against-5-indonesian-companies-over-haze/ (accessed 11 October 2020).

4 Selecting Digital Technologies to Address Development Objectives

When laying out DTs' ability to address the key issues at the heart of this publication (CCM, CCA, environmental sustainability, and DRM), this chapter reviews their current breadth and scale of application; and relevance to the key issues within climate, DRM, and environmental sustainability. This chapter then takes a deep dive into cost-effectiveness issues, laying out notably the key tools of Marginal Abatement Cost Curves and Marginal Adaptation Cost Curves which can form the basis of informed choices between DTs based on cost. This is followed by a discussion of user choices among DTs, security and privacy concerns, and the potential tensions between the introduction of DTs, on the one hand, and traditional knowledge and coping mechanisms, on the other hand. The chapter concludes with a discussion of the opportunities, enablers, and barriers in utilizing digital technologies.

Review of Current Breadth and Scale of Application

The discussion here attempts to provide a partial answer to the question of whether DTs can address these issues in their current scale, and their ability to be scaled further.

Concerning the current scale:

Stage I. DTs are well established, even in low-income communities of developing countries. What is needed on the part of development assistance institutions is support to communities in making content specific to CCM, CCA, DRM, and environmental sustainability issues, and to incentivize participation.

Stage II. DTs hold much potential in Asia given the high rates of usage and growth rates of these technologies in the region. In 2020, 65% of the Chinese population were active social media users,[155] and over 55% owned smartphones,[156] which enable their owners to use apps. Both numbers have been constantly growing over the last years and are expected to continue growing. Hence, development agencies should encourage the creation of social media and apps relevant to CCM, CCA, DRM, and environmental sustainability, for example, by sponsoring relevant creations in incubators and accelerators where private risk capital will not.

Stage III. Some Asian countries (the Republic of Korea, the PRC, Japan) are at the global forefront of Stage III DT. Regarding the PRC, a 2017 *Economist* report noted that the country could be a "close second" to or "even ahead of" the US in artificial intelligence (AI).[157] In a 2016 report, the US White House noted that the PRC had

---

[155] L. L. Thomala. 2021. Penetration rate of social media in China 2013–2020. *Statista*. 19 February. https://www.statista.com/statistics/234991/penetration-rate-of-social-media-in-china/#:~:text=In%202020%2C%20the%20social%20media,approximately%20931%20million%20active%20users.

[156] Statista. 2020. Smartphone penetration rate in China from 2015 to 2023. https://www.statista.com/statistics/321482/smartphone-user-penetration-in-china/ (accessed 25 November 2020).

[157] *Economist*. 2017. China may match or beat America in AI: Its deep pool of data may let it lead in artificial intelligence. 15 July. https://www.economist.com/news/business/21725018-its-deep-pool-data-may-let-it-lead-artificial-intelligence-china-may-match-or-beat-america.

overtaken the US in the number of published journal articles on deep learning (footnote 157). In 2017, PRC firms accounted for 99 of the 314 blockchain patents and 473 of the 649 AI patents filed with the World Intellectual Property Organization (WIPO).[158] The numbers of patents related to the two technologies filed by US companies were 92 for blockchain and 65 for AI.

But there are significant concerns with sharing their progress, given the strategic nature (and hence sensitivity) of some DT Stage III applications, and of intellectual property (IP)-related concerns (with governments not wanting to lose exclusive use of Stage III-related research and development [R&D] they sponsored). One option to nonetheless move forward with reaping the benefits of Stage III technologies for developing countries in Asia is for Stage III research relevant to key CCM, CCA, DRM, and environmental sustainability to be sponsored by dedicated trust funds from the Stage III DT "leadership" countries, based on public benefit calculations jointly made with ADB.

Relevance to Key Climate-Related, Disaster Risk Management, and Environmental Sustainability Issues in Asia

A second answer to the question of whether DT can address these issues is, for the CCM and CCA context, whether the DT in question is targeted at the key climate issues in regional member countries. This in turn hinges on the questions of

(i)   whether the DT in question can make a difference in sectors particularly impacted by climate change, and

(ii)   whether the DTs have so far emerged in relevant climate technology needs assessments.

For DRM, the key question is whether the DT are aligned with the regional DRM agenda and whether key regional DRM institutions like the Asian Disaster Preparedness Center have DT-relevant capacity at all three DT stages discussed in this publication.

For environmental sustainability, the core question is whether DT development across Asia is aligned with the environmental sustainability priority areas for the region, as outlined, for example, in ADB's OP3. These are (i) air, water, and land pollution; (ii) biodiversity conservation; (iii) ecosystem management; and (iv) marine litter and blue economy. Naturally, the key problem here is that many of the problems in these four issue areas represent a market failure, and DTs readily addressing them will not emerge naturally without adequate incentives. This requires either government (or ADB) support for the creation of ADB's addressing these market failures, or the creation of markets (e.g., for ecosystem services) within which DT-based services can then thrive.

Cost-Effectiveness, Affordability, and "Bang for Buck"

When selecting DTs for CCM, CCA, DRM, and environmental sustainability, one of the key parameters is to define how cost-effective the DT is for its purpose. While basic traditional analyses like cost–benefit analyses or cost-effectiveness might go some way in helping with these decisions, specific tools for the climate technology space have also emerged recently. Hence, in this section, both mitigation and adaptation tools for comparing cost-effectiveness between technologies, will be defined and discussed. While both tools have so far mostly been applied for hardware, the principles hold for DTs as well.

158   A. Peyton. 2019. China bosses blockchain and AI patents. *Fintech Futures*. 21 January. https://www.fintechfutures.com/2019/01/china-bosses-blockchain-and-ai-patents/.

Defining Cost-Effectiveness

One tool used in this cost–benefit (or cost-effectiveness) context is so far mostly used in the climate mitigation context, but easily applicable to adaptation, DRM, and environmental sustainability fields as well. This is the Marginal Abatement Cost Curve (MACC), which allows for comparison of the costs and emission reduction potential of different mitigation measures.[159]

The MACC is helpful in decision-making as it provides a simple way to identify which projects are the most cost-effective per unit of CO_2-equivalent abated and which options offer the greatest reduction potential. It is a visual representation of a group of listed GHG reduction projects, from the most cost-effective per ton of carbon to the least cost-effective.[160]

It is important to emphasize that MACCs can lead to sub-optimal decisions because it is a tool that compares the cost-effectiveness of various mitigation activities, but does not assess other variables. MACCs cannot properly reflect the variability and uncertainty of exogenous aspects, such as hydrology or demand growth, or represent temporary interactions between a mitigation measure and a sector, such as the effect of storage in the electricity system; or between multiple mitigation measures, for example, the capture of a market fraction by a technology. Therefore, the MACC should be understood as another policy input that allows for more in-depth analysis to update the curve on an ongoing basis, rather than making categorical decisions about it.[161]

Figure 4 shows an example of an MACC. Each rectangle is an independent technology, where the width is its abatement potential and the height the cost of the mitigation option, in $/t CO_2:

As shown in the figure, every technology is represented by a rectangle where the x-axis of the curve presents the abatement potential of each technology. Therefore, the wider the rectangle, the stronger the mitigation effect. The height of each rectangle presents its cost-effectiveness, and the technologies are shown from the most cost-effective (negative costs, therefore representing savings) to the least cost-effective (largest costs). The interpretation of the MACCs should be as follows:[162]

  (i)   When the cost-effectiveness is negative (savings), it is advisable to provide further analysis and technical support on assessing the costs and verifying its mitigation cost.
  (ii)   If there is a carbon price operating in the location where the technology will be operating, whenever the abatement cost is cheaper than paying the carbon tax, investments should be mandatory.
  (iii) Additional co-benefits of the technologies should be assessed in case the abatement costs are not sufficiently attractive, i.e., when the cost is close to the carbon price, or when the available information for calculating abatement costs has high uncertainty.

This tool presents a quantitative approach for comparing the costs of different DTs for addressing their mitigation cost-effectiveness. It is advisable to complement such analysis with other relevant attributes of the technologies for ideal resource allocation in technology prioritization.

[159] Low Emission Analysis Platform (LEAP). n.d. *Marginal Abatement Cost Curve (MACC) Summary Reports.* https://leap.sei.org/help/leap. htm#t=Views%2FMarginal_Abatement_Cost_Curve_(MACC)_Reports.htm.

[160] Compared to the business-as-usual situation.

[161] J. Ponz-Tienda et al. 2017. *Marginal abatement Cost Curves (MACC): unsolved issues, and alternative proposals.* DOI: 10.1007/978-3-319-54984-2_12. https://www.researchgate.net/publication/316753935_Marginal_abatement_Cost_Curves_MACC_unsolved_issues_and_alternative_proposals; P. Ekins, F. Kesicki and A.Z.P.Smith. 2011. *Marginal Abatement Cost Curves: A call for caution.* UCL Energy Institute. http://www.homepages.ucl.ac.uk/~ucft347/MACCCritGPUKFin.pdf.

[162] V. Eory et al. 2017. Marginal abatement cost curves for agricultural climate policy: State-of-the-art, lessons learnt and future potential. https://www.sciencedirect.com/science/article/abs/pii/S095965261830283X.

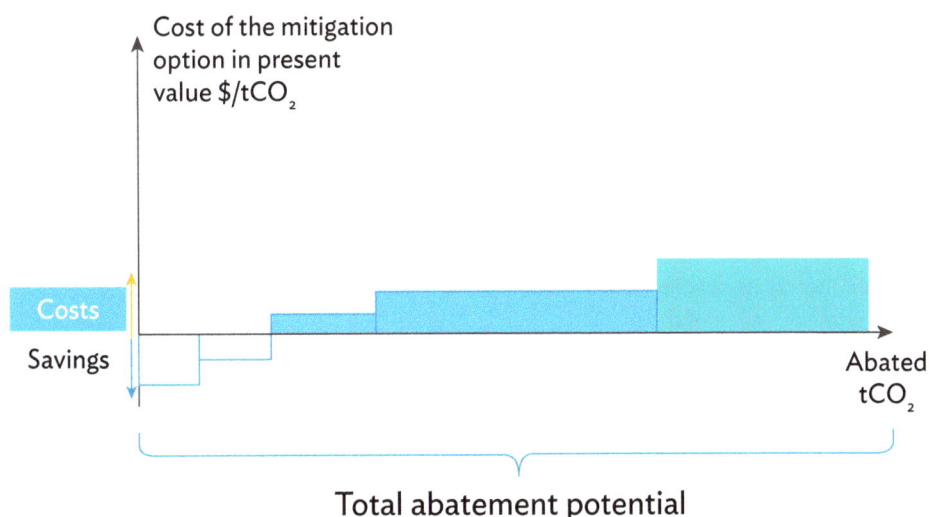

Figure 4: Example of Marginal Abatement Cost Curve

$/tCO_2$ = dollar per ton carbon dioxide.
Source: Created by and used with permission of ImplementaSur S.A. implementasur.com.

Defining Cost-Effectiveness in Adaptation Technologies

Similar to MACCs, Marginal Adaptation Cost Curves (MAdCCs) allow the comparison of the cost-effectiveness of various adaptation options. In contrast to mitigation, however, where a clearly defined metric of effectiveness (equivalents of total carbon dioxide [tCO_2] abated) allows a relatively simple ranking of alternative technologies, this is not the case in the adaptation domain.[163]

The diverse metrics of effectiveness and spatial and sector differentiation for abatement options present a challenge in establishing common metrics for assigning an order of merit in a MAdCC.[164] MAdCCs have the same shape as MACCs, ranking the selected adaptation technologies in terms of the total cost and the achieved degree of adaptation (footnote 163). The horizontal axis shows the achieved degree of adaptation, which is equal to the damage avoided by the implementation of the examined adaptation technologies or measures (footnote 163). The degree of adaptation is estimated by comparing the baseline scenario (where no implementation of adaptation technology or measures is assumed) with a scenario in which specific adaptation schemes are implemented (footnote 163).

The interpretation of MAdCCs is slightly different from MACCs and depends on the chosen adaptation threshold in a particular context and a particular choice of scenario.

---

[163] M. Skourtos, A. Kontogianni and C. Tourkolias. 2013. Report on the Estimated Cost of Adaptation Options Under Climate Uncertainty. The CLIMSAVE Project–Climate Change Integrated Assessment Methodology for Cross-Sectoral Adaptation and Vulnerability in Europe.

[164] A MAdCC uses the metric "achieved degree of adaptation," which represents the damage avoided by the implementation of the measure. This degree of adaptation is estimated by comparing the baseline scenario with a scenario in which specific adaptation schemes are implemented, so each measure will have particular merits depending on the climate threat and the vulnerability it reduces. Unlike the MACC curve that has a defined metric of effectiveness that is compared to cost (CO_2 abated), as mentioned above, the MAdCC does not, which makes the establishment of an unequivocal ranking difficult.

Defining Cost-Effectiveness in Disaster Risk Management and Environmental Sustainability

Unlike the tools of MACCs and MAdCCs, there is no unified model for cost-effectiveness in the DRM and environmental sustainability fields. DRM, being closely linked to risk management studies, evidently has seen a number of cost-effectiveness approaches being applied, but mostly tends to rely on cost–benefit analyses.[165]

Environmental sustainability is too broad a field for every cost-effectiveness approach to be analyzed here. Again, cost–benefit analyses are the fallback option for relevant cost-effectiveness calculations, though the breadth of the field has notably given rise to wide literature about the range of environmental sustainability benefits that may be attained (such as animal or plant species recovered, tons of emissions of a given pollutant, hectares of farmland preserved, kilometers of river length restored, and others), and methodological difficulties, for example, in defining reliable baselines in each of these.

Decision-Making Methods for Choosing among Digital Technologies

The cost-effectiveness of technologies, in terms of their abatement (or adaptation) potential and their costs, is a first step for assessing the applicability of the different DTs being analyzed for mitigation and adaptation. Nevertheless, it will be relevant to consider other criteria for determining how adequate are the technologies in question for the context where they will be inserted.

In this sense, the new market's reality in which the technology will be inserted and the user's perception of the "usability" of the solution is crucial to be considered. For these effects, a multi-criteria analysis (MCA) is a decision-making tool that can include additional qualitative criteria for ranking the technologies and determining how appropriate they are for the local context.

The MCA includes other parameters, including ease of local implementation, broader socio-environmental benefits, compatibility with current technologies, or the presence of local capacities for implementing technologies. One of the MCA methods most used is the Analytic Hierarchy Process (AHP) developed by Thomas L. Saaty, which is the selected methodology that will be briefly described below.[166]

It is suggested that a set of no more than nine criteria is designed and that these are defined with stakeholders relevant for the implementation of the DTs. The different criteria are then ranked in order of importance, and with formulas from the AHP methodology, a ponder is assigned to each, ranging from 0 to 1, summing 1 in total.

Ultimately, the technologies are evaluated considering each criterion, pondered by its relevance. Table 4 provides an example of a set of criteria that could be applicable for mitigation and adaptation DTs.

165   C.M. Shreve. 2014. Does mitigation save? Reviewing cost-benefit analyses of disaster risk reduction. *International Journal of Disaster Risk Reduction*. Vol. 10, Part A. pp. 213–235.

166   Department for Communities and Local Government: London. 2009. *Multi-criteria analysis: a manual*. January. http://eprints.lse.ac.uk/12761/1/Multi-criteria_Analysis.pdf.

Table 4: Example of a Set of Criteria for Prioritizing Adaptation and Mitigation Digital Technologies Using a Multi-Criteria Analysis, Along with the Priority Vector

Criteria defined with stakeholders for evaluating Digital Technologies	Priority Vector for Decision-Making (Relative Relevance between Criteria)	Weighing scale (1 to 5) that is assigned by an expert judgment for all criteria to a given technology. The score obtained in each criterion is corrected for the relative importance.				
		1	**2**	**3**	**4**	**5**
Size of GHG reduction or adaptation potential	0,1	Low GHG abatement or low degree of adaptation				Considerable GHG abatement or large degree of adaptation.
Enabling infrastructure	0,2	There is no enabling infrastructure available.				The DT is compatible with the current infrastructure.
Adequate time horizon	0,08	The technological option will hardly be able to adjust to the appropriate time horizons in the local context.				The technological option can be adjusted to an appropriate time horizon in the local context.
Cost-efficiency (MACCs and MAdCCs output)	0,06	The investment is very high in relation to the efficiency in reducing emissions /degree of adaptation.				The investment is low in relation to the efficiency in reducing emissions/ degree of adaptation.
Technology maturity	0,12	The technology is in the initial stages (theoretical, prototype)				The technology is in a commercial stage.
Initial capital required	0,15	The initial capital required is too high to be paid by the company. It will probably require a subsidy from the State to be feasible.				Companies can afford the initial investment for the technology option. No state subsidy should be required for this technology option.
Payback period	0,04	The payback period is high compared to other technology options.				The payback period is low compared to other technology options.
Existence of local capacities	0,25	There are no current local capacities for the implementation of the DT.				Similar experience in using other technologies has built capacities applicable to the new technology.

DT = digital technology, GHG = greenhouse gas, MACC = Marginal Abatement Cost Curve, MAdCC = Marginal Adaptation Cost Curve. Source: Created by and used with permission of ImplementaSur S.A. implementasur.com.

The AHP method utilizes an evaluation matrix that allows assigning a "final grade" to the different solutions, taking into account the priority vector and the individual score of the technology for each criterion. Finally, the ones with the highest score should be considered as the most adequate DTs for the local context,[167] integrating key parameters such as usability, technical complexity, and others.

---

[167] The evaluation is highly susceptible to the selected criteria. Therefore, it is necessary to include all relevant stakeholders in the definition of the parameters.

Security and Privacy Concerns

Security and privacy issues can be traced back to many of the DTs presented in this publication. The internet is home to a growing number of organizations that collect personal information from their users. Every user leaves traces that either directly or indirectly point to sensitive personal information (sometimes even from trivial information like internet shopping); social media interactions often lead to the sharing of sensitive personal information.

According to business insiders, 65 billion IoT devices are expected by 2025 (Figure 5). These technologies offer a wide range of possibilities, but also offer hackers more entry points, including critical infrastructures such as power grids. The prevalence of insecure IoT devices has been a concern. Such devices are estimated to account for 16% of traffic, but 78% of mobile malware.[168] Most IoT devices are prone to cyberattacks because not enough consideration is given to security in their design. For instance, it is not practical to install a firewall on most IoT devices due to insufficient memory.[169] The amount of data generated, which is at the heart of IoT, can leave sensitive information vulnerable.[170]

Digital technologies often pose risks to the privacy and security of their active and passive users because they require data and information for reliable operation or collect and produce data and information by themselves for further processes. This cannot completely be excluded as new technologies introduce new risks which are often

Figure 5: Amount of Internet of Things Devices Worldwide
(in billion)

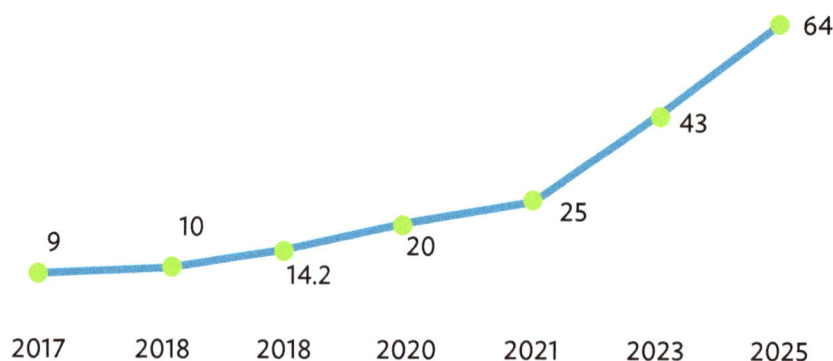

Sources: P. Newman. 2019. IoT Report: How Internet of Things technology growth is reaching mainstream companies and consumers. *Business Insider*. https://www.businessinsider.com/internet-of-things-report; G. Omale. 2018. Gartner Identifies Top 10 Strategic IoT Technologies and Trends. Gartner. https://www.gartner.com/en/newsroom/press-releases/2018-11-07-gartner-identifies-top-10-strategic-iot-technologies-and-trends, https://www.gartner.com/en/newsroom/press-releases/2018-11-07-gartner-identifies-top-10-strategic-iot-technologies-and-trends; Connected Devices: Microsoft's Heritage of Innovation. Microsoft. https://www.microsoft.com/en-us/legal/intellectualproperty/mtl/internet-of-things.aspx; F. Dahlqvist et al. 2019. Growing opportunities in the Internet of Things. McKinsey & Company. https://www.mckinsey.com/industries/private-equity-and-principal-investors/our-insights/growing-opportunities-in-the-internet-of-things.

[168]   T. Mann. 2019. Nokia VP: 5G Security Risks Are Huge. SdxCentral. https://www.sdxcentral.com/articles/news/nokia-vp-5g-security-risks-are-huge/2019/10/.

[169]   D. Gayle. 2016. Smart' devices 'too dumb' to fend off cyber-attacks, say experts. *The Guardian*. 22 October. https://www.theguardian.com/technology/2016/oct/22/smart-devices-too-dumb-to-fend-off-cyber-attacks-say-experts].

[170]   *Business Insider*. 2020. The security and privacy issues that come with the Internet of Things. 7 January. www.businessinsider.com/iot-security-privacy?r=DE&IR=T.

accepted to come up with innovations faster.[171] However, organizations have to find the right balance between innovation and security in digital transformation. A start can be to try to fully understand their cyber-exposure across all applied digital technologies.[172]

Traditional Knowledge and Avoidance of "Technology Idolatry"

This publication should also mention the instances where DTs might provide the most effective solution to a CCM, CCA, DRM, or environmental sustainability. There is much recent literature about the careful approach needed when deploying DTs in this context.[173] Some key lessons found include to not consider DTs as a panacea in this context, and when DT deployment is decided upon, emphasize knowledge components; accessibility (e.g., use of local languages); and local design components. The process of DT deployment also matters, with preference to be given to interactive processes. Finally, when designing DT interventions in this context, impact assessments may help elucidate unintended side-effects such as the displacement of traditional local coping mechanisms for the problem at hand.

Opportunities, Enablers, and Barriers in Utilizing Digital Technologies

Opportunities and Enablers

The use of DTs to promote environmental sustainability by companies in ADB DMCs can offer opportunities to develop competitive advantages to successfully exploit such advantages in foreign markets, especially in European Union (EU) economies. For instance, in July 2020, the European Commission (EC) launched public consultations on taxation rules to meet the EU's climate goals.[174] The public consultation document intended to revise the internal Energy Tax Directive by introducing a carbon border adjustment mechanism. The idea is to reduce carbon leakage risks and discourage companies from shifting their production activities to countries that have less stringent green regulations and poor enforcement mechanisms.[175]

Some companies are proactively taking steps to ensure that their supply chains are ready to comply with new regulations (footnote 78). With the debate about legislation aimed toward reducing carbon leakage risks and the possibility of carbon dioxide (CO_2) taxes in EU economies in the near future, organizations that have systems in place to calculate and track their CO_2 footprints will be at an advantage.[176] In this way, blockchain or distributed database solutions offer a proactive way to manage regulatory demands and the impact of new regulations (footnote 78).

To give an example, the carmaker Volvo's goal has been to generate 50% of its global car sales from fully electric models by 2025. It also aims to achieve a 50% reduction in tailpipe carbon emissions per car between 2018 and

[171] Capgemini. 2019. *Digital Trust & Security - Securing Digital Transformation.* https://www.capgemini.com/de-de/wp-content/uploads/sites/5/2019/10/Securing-Digital-Transformation-Capgemini-Invent.pdf.

[172] B. Egner. 2018. *How to Balance Security with Digital Transformation.* https://www.infosecurity-magazine.com/opinions/balance-security-digital/.

[173] G. Brodnig et al. 2017. Bridging the Gap: The Role of Spatial Information Technologies in the Integration of Traditional Environmental Knowledge and Western Science. *Information Systems in Developing Countries.* Vol 1 (1).

[174] EU Commission Press. 2020. Taxation and the European Green Deal: Commission launches public consultations on Energy Taxation Directive and a Carbon Border Adjustment Mechanism, EU Commission Press. *PubAffairs Bruxelles.* https://ec.europa.eu/taxation_customs/news/commission-launches-public-consultations-energy-taxation-and-carbon-border-adjustment-mechanism_en.

[175] I. Todorović. 2020. EU preparing CO_2 tax for products from other countries. *Balkan Green Energy News.* https://balkangreenenergynews.com/eu-preparing-co2-tax-for-products-from-other-countries/.

[176] N. Gibbs. 2020. Automakers call out industry's weak spots: Companies seek more ethical, resilient supply chains *Automotive News. Detroit.* 94 (6947). p. 8. https://www.autonews.com/suppliers/automakers-seek-more-ethical-resilient-supply-chains.

2025.[177] The carmaker has embraced blockchain to achieve these goals and to demonstrate the environmental and social responsibility of its activities (footnote 78).

Another carmaker, Mercedes-Benz, has also explored blockchain's potential to promote sustainability with a primary focus on environmental sustainability (footnote 78). Mercedes-Benz, a division of Daimler, teamed up with blockchain company Circulor to conduct a pilot project that involves the use of blockchain to track CO_2 emissions in the supply chains of its battery cell manufacturers.[178] It also tracks secondary materials, which are materials that are used, recycled, and sold for use in manufacturing. The goal is also to document whether Daimler's sustainability standards are passed on throughout the supply chain (footnote 178). A blockchain-based system records the production flow of the materials and CO_2 emissions. It also records the amount of recycled material in a supply chain. The network also displays working conditions, environmental protection, safety, business ethics, compliance, and human rights (footnote 178). The company's goal is to evaluate whether these indicators meet Daimler's sustainability requirements. Daimler will ask its direct suppliers to comply with the relevant standards. Upstream value chains are also expected to comply. Pressures can be passed down to upstream value chains.

A related point is that small companies from ADB DMCs that fail to develop such capabilities may not be able to do business with global multinational enterprises. It was reported that Volvo discontinued at least one supplier from its network due to noncompliance with its demand to connect to blockchain networks and share data related to CO_2 emission (footnote 78). Volvo plans to expand its monitoring to heavy components to reduce CO_2 emissions throughout its supply chain network.[179] Thus, in such cases, being part of blockchain network or any other IT system mandated by Volvo is matter of a firm's survival (footnote 78).

Education and awareness on navigating the universe of DTs successfully is a key enabler for the successful use of DTs in any domain. A number of governments have, therefore, made considerable efforts to integrate learning on and with DTs into national curricula, covering in most cases the design and development of digital "outcomes" (like digital media) and more in-depth computational thinking (programming).[180] Evidently, this means fundamental changes to previous training and practice for teachers.[181] The Organisation of Economic Co-operation and Development (OECD) Programme for International Student Assessment (PISA), possibly the most authoritative evaluation of educational systems by scholastic performance, has for this reason recently commissioned a study on the comprehensive integration of DTs in schools.[182]

A key enabler for those who are already users of various forms of DT are digital platforms for their effective use. Early-stage platforms often focus on experience exchange among users or trouble-shooting for new DTs. As users get more comfortable and increasingly push to maximize the technologies they are using, many improve on existing features or suggest new technologies. Ultimately, such platforms may push the boundary of idea generation and usher in new R&D,[183] with relevant alliances and networks emerging.[184]

[177] J. Nerad. 2020. Volvo Set to Challenge Tesla for Electric Car Supremacy. *Forbes*. https://www.forbes.com/sites/jacknerad2/2020/01/22/volvo-set-to-challenge-tesla-for-electric-car-supremacy/#613480e34011.

[178] J. Nastu. 2020. Mercedes-Benz Pilots Blockchain Project to Track Emissions throughout Supply Chain. *Environment + Energy Leader*. https://www.environmentalleader.com/2020/01/mercedes-benz-pilots-blockchain-project-to-track-emissions-throughout-supply-chain/.

[179] P. Campbell. 2020. EV supply chains seek clearer visibility with blockchain. *Financial Times*. https://www.ft.com/content/3652b68e-206f-4e3a-9b3a-4d5ec9a285b7.

[180] For example, the courses provided by "Code Avengers" for the New Zealand Ministry of Education classify their approaches in this way. Government of New Zealand, Ministry of Education. *Code Avengers*. https://www.codeavengers.com/profile#all (accessed 11 October 2020).

[181] C. Blundell, Christopher, K.-T. Lee, and S. Nykvist. 2016. Digital learning in schools: Conceptualizing the challenges and influences on teacher practice. *Journal of Information Technology Education:Research*, 15. pp. 535–560.

[182] OECD. 2019. PISA 2021. ICT Framework.

[183] M. Hossain et al. 2017. How do digital platforms for ideas, technologies, and knowledge transfer act as enablers for digital transformation? *Technology Innovation Management Review*, Vol. 7 (9). September 2017. pp. 55–60.

[184] J. West and M. Bogers. 2014. Leveraging external sources of innovation: a review of research on open innovation. *Journal of Product Innovation Management*, 31 (4).

Barriers and Strategies for Overcoming Them

One of the major barriers is the lack of availability of basic DTs. For instance, in countries such as Afghanistan and Pakistan, a large proportion of the population is not covered by the mobile-cellular network or the internet (Figure 6).

Figure 6: Internet Users per 100 People in Selected Asian Countries

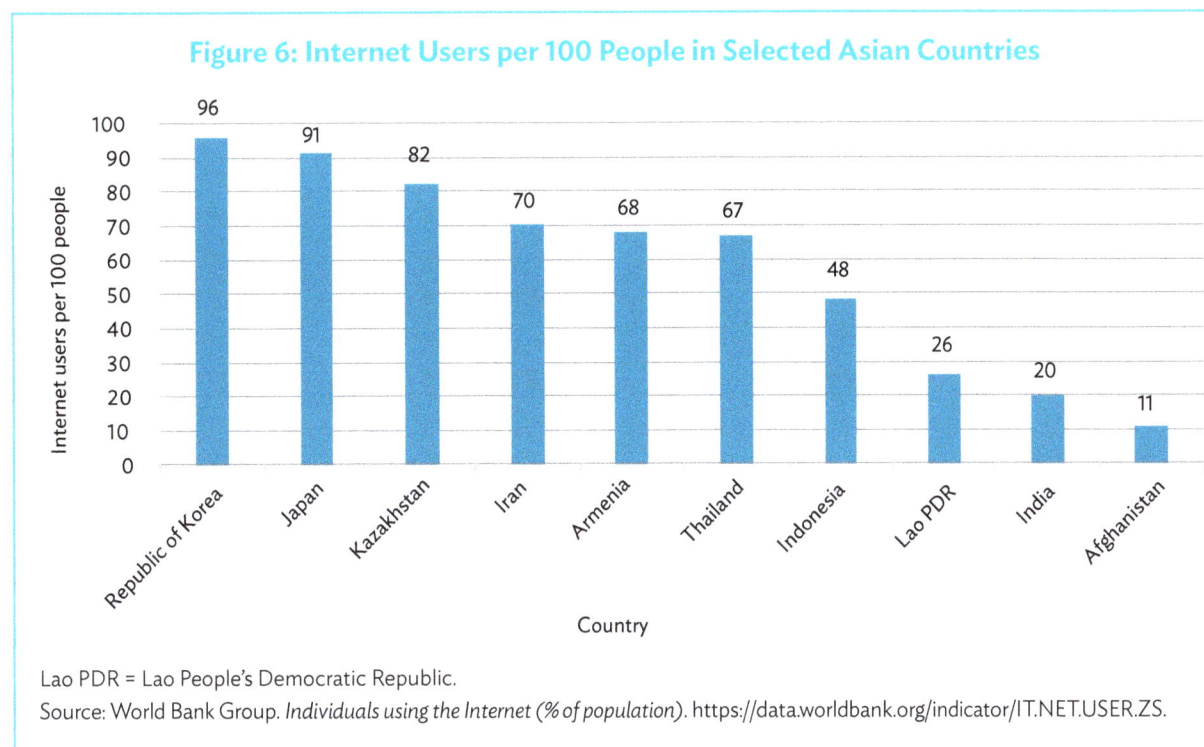

Lao PDR = Lao People's Democratic Republic.
Source: World Bank Group. *Individuals using the Internet (% of population)*. https://data.worldbank.org/indicator/IT.NET.USER.ZS.

Where mobile phones or internet connectivity exists, it is frequently unreliable. Basic technical barriers are non-existent or unreliable electricity, or slow internet speeds (Figure 7).[185]

Further issues include illiteracy, lack of capacity to pay for DTs or DT-related services, or lack of locally appropriate content. Other basic barriers are regulatory, with some governments in the region concerned notably about DTs that enable communication or collaboration at odds with the government, or with the country's legal systems, and a tightening of cybersecurity laws most recently in the PRC.[186] This has obviously also spilled over into the most advanced technologies, with, for example, the PRC promulgating a 2019 law that regulates the use of participants' identities in blockchain,[187] a move seen by many analysts as likely hampering certain types of blockchain activity in the PRC. Evidently, there is a great level of international heterogeneity in how countries approach regulation of blockchain activity, with the most permissive regulations mostly in the EU, the US, Australia, and New Zealand.[188]

---

[185]   Fastmetrics. n.d. *Average Internet Speeds By Country*. https://www.fastmetrics.com/internet-connection-speed-by-country.php (accessed 25 November 2020).
[186]   Hogan Lovells. 2018. Asia Data Protection and Cybersecurity Guide 2018.
[187]   Ledger Insights. 2018. *China publishes blockchain service, identity regulations*. https://www.ledgerinsights.com/china-blockchain-regulations-identity-censorship/.
[188]   U.W. Cohan. 2017. Assessing the Differences in Bitcoin & Other Cryptocurrency Legality Across National Jurisdictions. *SSRN* (revised 2 April 2020). https://papers.ssrn.com/sol3/papers.cfm?abstract_id=3042248.

Figure 7: Internet Speed in Selected Asian Countries

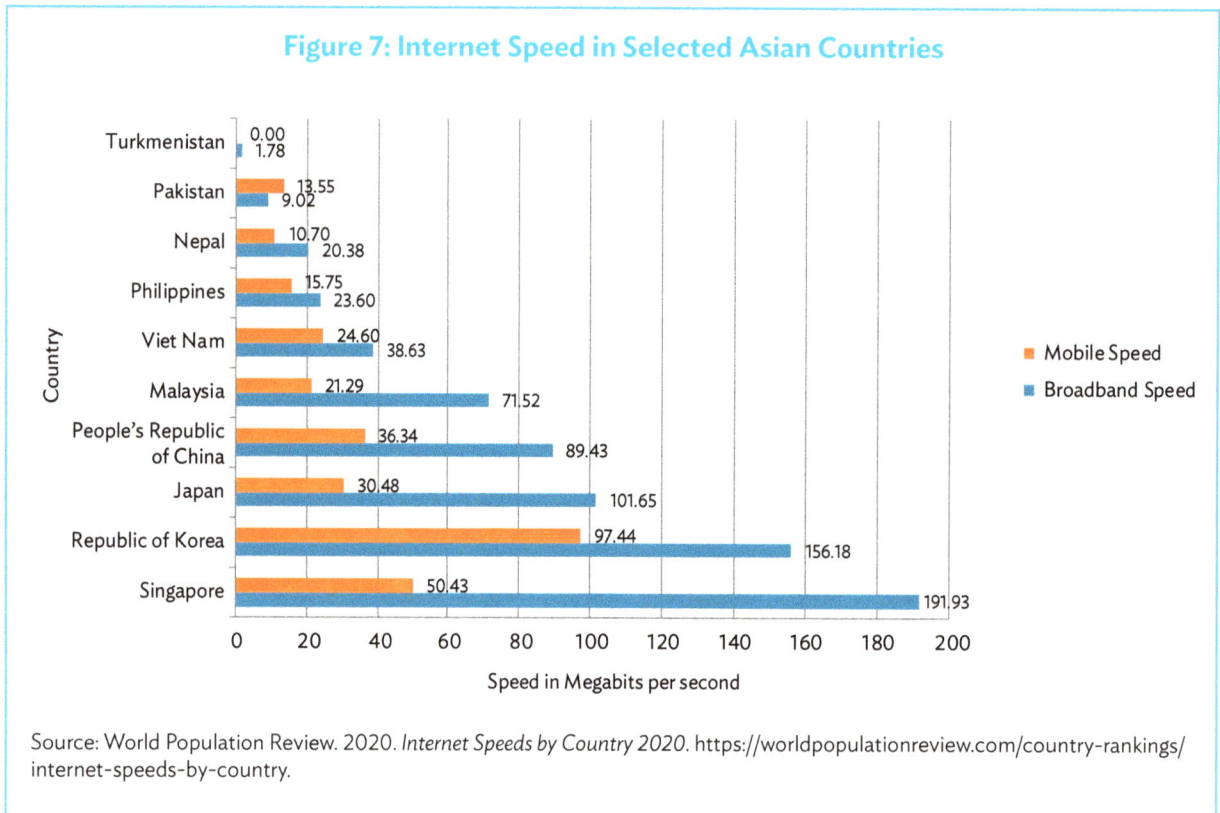

Source: World Population Review. 2020. *Internet Speeds by Country 2020*. https://worldpopulationreview.com/country-rankings/internet-speeds-by-country.

Besides, DTs face technology-specific "scaling" barriers, such as

(i)   The ability to scale the sourcing of relevant raw materials for the type of DT in question. This is, of course, a cost and logistics issue, but also leads to the environmental and resource depletion concerns laid out in section 3.1, notably for Stage I DTs.

(ii) The ability to muster the additional human resources required—particularly difficult for DTs from Stage III, which require highly skilled individuals to run them effectively. For instance, according to a study conducted by Tencent Research Institute, there were only about 300,000 "AI researchers and practitioners" in the world in 2017.[189] There is a demand for millions of such roles in companies and government agencies all over the world. There is even a larger gap between demand and supply for higher quality AI workforce. According to Montreal-based lab, Element AI, the entire world has fewer than 10,000 people that have serious AI research skills.[190] According to the Tencent report, the US, the PRC, Japan, and the United Kingdom (UK) account for most of the AI workforce. The next two countries with the concentration of AI workforce are Israel and Canada. Among these countries, the US is reported to be far ahead of others. A main reason is that the US has the highest number of universities that offer ML-related courses. The US also has more AI start-ups compared to any other nations. The report estimated that there were 2,600 AI start-ups worldwide of which the US had more than 1,000 and the PRC had about 600 (footnote 191).

---

[189] J. Vincent. 2017. Tencent says there are only 300,000 AI engineers worldwide, but millions are needed. *The Verge*. https://www.theverge.com/2017/12/5/16737224/global-ai-talent-shortfall-tencent-report.

[190] C. Metz. 2017. Tech Giants Are Paying Huge Salaries for Scarce A.I. Talent. *The New York Times*. https://www.nytimes.com/2017/10/22/technology/artificial-intelligence-experts-salaries.html.

Figure 8: Percentage of the Population Covered by a Mobile-Cellular Network in Selected ADB Members, 2018

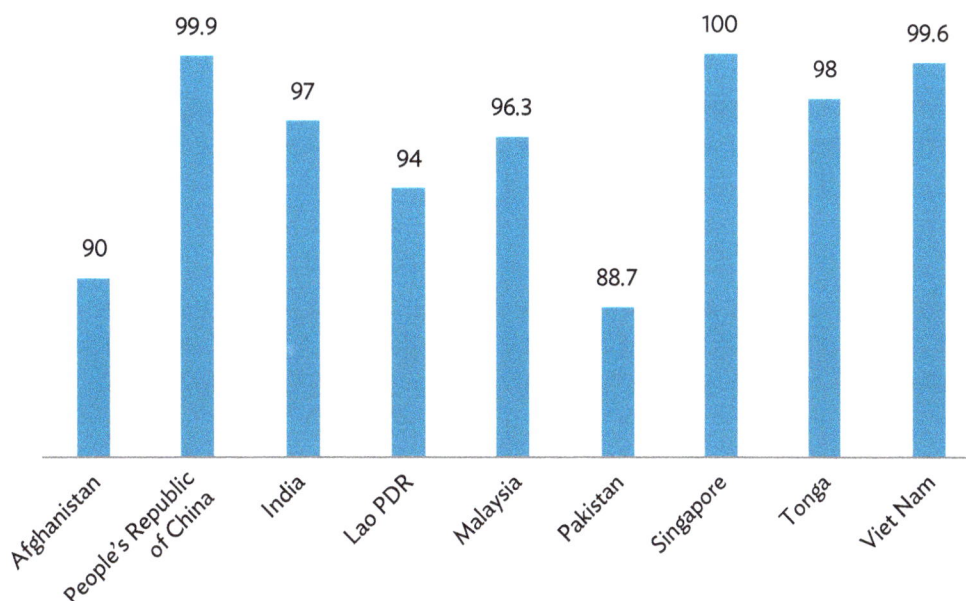

Country	%
Afghanistan	90
People's Republic of China	99.9
India	97
Lao PDR	94
Malaysia	96.3
Pakistan	88.7
Singapore	100
Tonga	98
Viet Nam	99.6

Lao PDR = Lao People's Democratic Republic.
Source: World Bank Group. *Individuals using the Internet (% of population).* https://data.worldbank.org/indicator/IT.NET.USER.ZS.

There are, of course, a range of possible strategies that can be employed to overcome these barriers. Key among these are DT-specific technology needs assessments in climate-relevant sectors, and quantitative market analysis for DTs in climate- and sustainability relevant sectors. Key points to be analyzed are size of a market, transaction costs, or even, for example, the likelihood of network effects emerging for DTs where this is needed for the DT to fulfill this function (e.g., network effects are crucial for many apps to function effectively).[191]

A lack of appropriate regulatory and legal structures, including an incentive to use advanced DTs and bureaucratic resistance to do so requires the building of such structures and of the institutions that can oversee and enforce them. These institutions should also be in charge of adapting DT strategies to local circumstances. This will require local innovation environments that can lead this technology adaptation process based on DT invented elsewhere, or that can build needed indigenous DT.

---

[191] "Network effects" means that increased numbers of users improve the value of the good or service.

5 Operationalizing Digital Technologies in Developing Asia

How can DTs help address the key challenges faced by ADB DMCs in the climate change, DRM, and environmental sustainability arena? On the climate change mitigation front, ADB DMCs obviously cover the entire range of GHG emission mitigation challenges, with a recent International Monetary Fund (IMF) publication notably highlighting rising energy use and vanishing land cover.[192] As for climate change adaptation, the Fifth Assessment Report of the IPCC notably highlights water scarcity and food security.[193] As for DRM, a 2019 priority ranking by UNESCAP lists earthquakes and storms, followed by floods.[194] Finally, regarding environmental sustainability, recent studies highlight unsustainable resource management and natural resource depletion, ecosystem degradation and biodiversity loss, pollution, and waste.[195]

Maximizing the Use of Digital Technologies from Each Stage to Specific Stakeholder Groups

Table 5 summarizes how various stakeholders such as consumers, companies, activists, nongovernment organizations (NGOs), and government agencies can use various DTs discussed above to achieve various goals related to CCM, CCA, DRM, and environmental sustainability. In the cells of the resulting matrix, it discusses at least one example from each stage.

Stage I DTs above represent a treasure trove of basic information, search, and communication possibilities. The key challenges here are threefold: (i) to make the information easily accessible and understandable, and to enable information recipients to undertake concrete actions in climate change, environmental sustainability, and DRM space; (ii) to make search functions easy to use, yielding information targeted at the individuals, entities, or groups searching for them; and (iii) to make communication possibilities easy to embed in coordinated efforts in climate change, environmental sustainability, and the DRM space.

As for Stage II DT, well-designed apps and social media that have proven useful to a large enough audience will become self-perpetuating. Designers of relevant apps and social media in climate change, environmental sustainability, and the DRM space, therefore, have two tasks: (i) to support the creation of relevant climate and sustainability social media and apps that the market will not bring about on its own; (ii) to prioritize standardization and quality control efforts.

As is evident from the preceding discussion on Stage III, this set of technologies is the most complex and costly. This makes structured technology choice mechanisms crucial. Therefore, considerable time was allotted on such mechanisms, including MACCs and multi-criteria analysis.

---

[192]   A. Prakash. 2018. Boiling Point. *Finance and Development*. IMF.
[193]   Intergovernmental Panel on Climate Change. 2014. *Climate Change 2014: Impacts, Adaptation, and Vulnerability*. Part B: Regional Aspects.
[194]   UNESCAP. 2019. Summary of the Asia-Pacific Disaster Report 2019.
[195]   UNESCAP. 2018. Key environment issues, trends and challenges in the Asia-Pacific region.

Table 6: How Stakeholders Can Use Digital Technologies in Climate Change Mitigation, Climate Change Adaptation, Disaster Risk Management, and Environmental Sustainability

Stakeholder	Mitigation	Adaptation	Disaster Risk Management	Environmental Sustainability
Consumers	• Using DTs, such as **IoT and Smart Systems** to improve the energy efficiency in buildings. Different technologies, such as smart heating, air-conditioning, lighting, and automated systems can be used to optimize energy distribution, automate building operations, and improve performance.[a] (Stage III)	• **5G technologies** can be used to reduce water losses in areas affected by climate-induced water shortage, by enabling real-time sensors and analysis.[b] (Stage III)	• Use of **smartphone apps** to assist the response to disasters. Collecting data from social media to assess situations and interactively use the crowd as reporters.[c] (Stage II)	• Use **apps** to promote and offer services for sustainable mobility alternatives or provide feedback about current mobility behavior. They can also be used to monitor energy consumption in order to provide information that could help energy efficiency in buildings.[d] (Stage II) • These apps can be used as basis to understand **Big Data** and harness **AI** to create practical applications that can be scaled by non-technical users. (Stage III)
Companies	• Using DTs to improve data center management. ICTs and **AI technologies**-based solutions can be used to control a data center's cooling facilities, optimizing energy consumption. (Stage III)	• Using mobile **apps** to enhance food security and support the agriculture sector, by delivering weather and climate information, use DTs for smart agricultural production, and for micro-level drought preparedness. (Stage II)	• Crisis analysis and visualization tool that provides real-time situational information from various data sources to enhance disaster management efforts. Using information from **satellites** and other sources to locate hazards and provide response strategies. (Stage I)	• Using DTs to monitor food security, water transportation, and supply. The DTs that can be used for this purpose include **satellites**, Global Positioning Systems (**GPS**), and machine-to-machine connectivity that supports **remote sensing.** infrastructure. (Stage I)

continued on next page

Table 6 *continued*

Stakeholder	Mitigation	Adaptation	Disaster Risk Management	Environmental Sustainability
Nongovernment organizations	• Deployment of **drones** to build 3D models of forest surfaces. By capturing information about tree health, height, biomass, and other factors to provide estimates of the amount of carbon storage, the drones provide unparalleled access to highly precise information about carbon capture that is used to determine long-term implications for scientific research, management, conservation, monitoring, and other uses. (Stage III)	• Using DTs for climate-related humanitarian action ("climate refugees") by gathering critical information, improve their internal services, and better support those in need. For example, the usage of **mobile phone data** to assess climate change and migration patterns. (Stage I)	• Use of **blockchain** for a more rapid and reliable collection of data during a crisis.	• Using DTs to **monitor** the global environment and ecosystem, such as weather **satellites** that track the progress of hurricanes and typhoons, weather radars that track the progress of tornadoes, thunderstorms, and major forest fires, and earth observation satellite systems that obtain (atmospheric) environmental information. (Stage I)
Government	• Using **smart grids** to reduce energy demand and accelerate renewable uptake among the population. Smart grids also benefit distributed renewable generation by providing real-time information on the performance of the system. (Stage III)	• Using DTs to establish early warning systems (EWSs) and enhance disaster management to **circulate weather information**. With this information, cities can deliver early warnings to communities at risk of climate impact events. (Stage III)	• Use of **AI and ML** to recognize damaged buildings, flooding, impassable roads, and others from satellite photos. Crowdsourced data analysis and ML can also be used to identify locations affected by hazards that had not yet been assessed or had not received aid. (Stage III)	• Using DTs to monitor deforestation and forest degradation. **Satellites** that are now able to take images through clouds (and at night) and **remote sensing** applications are critical for monitoring the health of forests. (Stage I)

5G = fifth generation, AI = artificial intelligence, ICT = information and communications technology, DT = digital technology, IoT = Internet of Things, ML = machine learning.

a   ITU. 2019. *Turning digital technology innovation into climate action*. https://www.uncclearn.org/wp-content/uploads/library/19-00405e-turning-digital-technology-innovation.pdf.
b   ITU. 2020. *Frontier technologies to protect the environment and tackle climate change*. Geneva, Switzerland. https://www.itu.int/en/action/environment-and-climate-change/Documents/frontier-technologies-to-protect-the-environment-and-tackle-climate-change.pdf.
c   ITU. 2019. *Disruptive technologies and their use in disaster risk reduction and management*. https://www.itu.int/en/ITU-D/Emergency-Telecommunications/Documents/2019/GET_2019/Disruptive-Technologies.pdf.
d   B. Brauer. et al. 2016. *Green By App: The Contribution of Mobile Applications to Environmental Sustainability*. https://www.researchgate.net/publication/308708034_GREEN_BY_APP_THE_CONTRIBUTION_OF_MOBILE_APPLICATIONS_TO_ENVIRONMENTAL_SUSTAINABILITY.
e   T. Brady. 2019. How emerging tech can counteract climate change. https://www.greenbiz.com/article/how-emerging-tech-can-counteract-climate-change.

Public–Private Partnership Projects

Public–private partnership (PPP) projects may be implemented to address some of the key challenges. For instance, social media companies and governments can collaborate to monitor social media contents to identify activities that can harm the environment and delicate ecosystems, and deplete resources. Illegal wildlife traders, especially operating in some countries such as the Lao People's Democratic Republic, Viet Nam, the PRC, and Thailand are reported to be using social media such as WeChat and Facebook for their criminal activities.[196] Tencent, which owns WeChat, has said that it utilizes big data to help catch ivory smuggling criminal syndicates. As of December 2015, Tencent reported that it closed 622 private accounts allegedly operated by illegal wildlife traders.[197] However, activists have argued that there is not enough policing in WeChat to deal with the spike in Illegal wildlife trading. In 2018, social media companies such as Facebook, Microsoft, and Tencent joined the Coalition to End Wildlife Trafficking Online, which aimed to reduce wildlife trafficking online by 80% by 2020.[198] Critics have been concerned that social media platforms lack firm commitments and sufficient resources have not been put into such efforts (footnote 197). PPP initiatives need to be carried out to fight such activities. Law enforcement agencies and social media companies should work in close collaboration and coordination to address these issues.

Increasing the Absorptive Capacity of ADB Developing Member Countries and Promoting Knowledge Sharing

Increasing the absorptive capacity of ADB DMCs is especially important for Stage III technologies such as AI. The lack of relevant skills is among the most serious challenges facing ADB DMCs. Finding high-quality AI talents such as ML engineers has been a big challenge for companies in these economies.[199] For instance, as of 2019, India was estimated to have about 50–75 AI researchers (footnote 141). According to Aspiring Minds' Annual Employability Survey 2019, only 2.5% of Indian engineers had AI skills required by the industry (footnote 142). India has also faced a severe shortage of qualified faculty members to teach AI courses in its universities (footnote 143).

It is also important to promote knowledge sharing to effectively utilize DTs. Some initiatives on this front in collaboration with multilateral agencies have already taken place. In April 2018, the United Nations Development Programme (UNDP) and the Chinese Academy of Governance (CAG) organized the "Workshop on South–South Cooperation under the Belt and Road Initiative: China's South–South Assistance to Disaster Recovery Efforts."[200] A main goal of the workshop was to share the PRC's expertise and technology in disaster recovery with other developing economies. Due to the PRC's vulnerability to natural hazards such as earthquakes, floods, droughts, and typhoons, it has developed a high degree of expertise in disaster recovery technologies. For instance, the Natural Disaster Reduction Center of China (NDRCC) has state-of-the-art technologies and methods to monitor natural hazards and disaster risk. It uses advanced tools such as imagery from orbiting satellites and drones. The NDRCC has teamed up with the United Nations Platform for Space-Based Information for Disaster Management and Emergency Response (UN-SPIDER) and the United Nations Office for Disaster Risk Reduction (UNDRR) to make the maps and satellite imagery available to other economies.[201] More initiatives like these need to be planned to achieve the goals of OP3.

---

[196] P. Yeung. 2019. How China's WeChat became a grim heart of illegal animal trading. *Wired*. 11 March.

[197] AllAfrica. 2015. Africa: Chinese Internet Giant Tencent to Promote Wildlife Conservation in Kenya. 16 December. http://allafrica.com/stories/201512160323.html (accessed 10 October 2020).

[198] World Wildlife Fund. 2018. Coalition to End Wildlife Trafficking Online. https://www.worldwildlife.org/pages/coalition-to-end-wildlife-trafficking-online.

[199] S. Sen. 2018. India moves to address AI talent supply gap, gets a leg-up from Google, Microsoft, Intel. *Factor Daily*. 18 January. https://factordaily.com/india-ai-talent-gap-google-microsoft/.

[200] UNDP. 2018. *Sharing China's Experience to Build Back Better*. https://www.cn.undp.org/content/china/en/home/presscenter/pressreleases/2018/sharing-china_s-experience-to-build-back-better.html.

[201] UNOOSA. 2019. Ten years of the UN-SPIDER Beijing office. Vienna. https://www.unoosa.org/res/oosadoc/data/documents/2019/stspace/stspace_0_html/19-07423_UN_SPIDER_ebook_spreads.pdf.

Integrating Digital Technologies into Nationally Determined Contributions

This section highlights the opportunities for DMCs to use DTs in meeting their own climate goals under their respective NDCs and raising ambition over time. The most straightforward role of DTs on this topic is in the area of monitoring, reporting, and verification (MRV) of emission reduction efforts. This will provide efficient and cost-effective certainty to the authority and to carbon markets and, therefore, will reduce transaction costs (since less validation will be needed). Also, DTs can help to track upstream GHG emissions from strategic products and commodities, to properly inform the market and investors.

Article 4.2 of the Paris Agreement requires Parties to include a mitigation contribution in their NDCs, as the most explicit provision on the component of NDCs.[202] Recent calculations reveal massive mitigation potential from doing so.[203] On the adaptation front, the Lima Call for Action invited Parties to consider communicating their undertakings in adaptation planning or including an adaptation component in their intended nationally determined contributions (INDCs).[204] Together, through its Article 3, the Paris Agreement states: "As nationally determined contributions to the global response to climate change, all Parties are to undertake and communicate ambitious efforts as defined in Articles 4, 7, 9, 10, 11 and 13 with the view to achieving the purpose of this Agreement as set out in Article 2.

The efforts of all Parties will represent a progression over time while recognizing the need to support developing country Parties for the effective implementation of this Agreement." Therefore, this article may be interpreted to mean that NDCs may contain components related to (i) mitigation and (ii) adaptation, covered extensively in this publication; (iii) financial support, which implies specific financial support to the scaling of DTs to climate ends; (iv) technology transfer, which means integrating DTs into climate technology needs assessments and corresponding financing requests; (v) capacity building, which requires integrating DT-specific capacity building into NDCs; and (vi) transparency, which means using DTs for transparency purposes (e.g., DLT).

Moreover, Article 4 of the Paris Agreement lists several characteristics required of NDCs, but does not refer to specific features as such. NDCs are required to, for instance, represent a progression from previous NDCs, which means this should usher in a progressive introduction of more complex DTs and DT applications in progressive NDCs. Secondly, NDCs are required to represent the highest possible ambition (and, whenever required, be adjusted for that purpose), which implies a prioritization of Stage III DTs over Stage II and Stage II over Stage I. Finally, they should be accounted for to promote environmental integrity; transparency; and ensure accuracy, completeness, comparability, and consistency, and avoid double counting (footnote 202). This means that DTs should correspondingly be used in NDCs for transparency; and data. More precisely, they should be used to build relevant databases under technologies of Stage I, data exchange under technologies of Stage II, and data transparency under technologies of Stage III.

---

[202] F.Z. Taibi and S. Konrad. 2018. Pocket Guide to NDCs under the UNFCCC. https://pubs.iied.org/sites/default/files/pdfs/migrate/G04320.pdf.

[203] For example, one recent study argues above all for massive abatement potential in digitization of the power sector, across both supply and demand intervention points. In the case of the PRC, for example, it argues that the digitally enabled accelerated decarbonization abatement of 777 metric tons of carbon dioxide equivalent ($MtCO_2e$) is equivalent to decommissioning over 170 average PRC coal-fired power plants. See GeSI. 2020. Digital Solutions for Supporting the Achievement of the NDC.

[204] UNFCCC. 2014. Lima Call for Action. https://unfccc.int/files/meetings/lima_dec_2014/application/pdf/auv_cop20_lima_call_for_climate_action.pdf.

Leveraging the Green Recovery

Green Recovery packages in light of COVID-19 can be leveraged toward greater integration of DTs into climate change, DRM, and environmental sustainability work. One of the main cornerstones of recommended policy measures to take action toward a green recovery is to focus on green investments by creating green stimulus packages. In this sense, the following mechanisms emerge as targets where public investment can be focused and at the same time offer a promising opportunity to seize the advantages of digital technologies:[205]

(i)   leveraging the climate-smart infrastructure;
(ii)   conditioning aid to the private sector for climate progress; and
(iii)   avoiding the prioritization of high-emission, high-risk, short-term projects.

These mechanisms can be applied with a strong DT lens to different sectors through the following recommendations for three key sectors that represent key investments of any green recovery:

Communications networks evidently have a big role to play in digitalizing the green recovery, given the need for social distancing and, at the same time, for continuity of social interactions and business operations. In this sense, countries will need to develop their digital connectivity infrastructure to improve the global experience of teleworking and social interaction.[206] The increase in traffic through the networks implies the need for investments in improving the networks to be able to meet the capabilities of the current and future scenarios. However, to achieve substantial and sustainable changes, an investment must also be made in reducing the environmental footprint of digital technologies.[207]

Modernization of the **electricity network** to make electricity systems resilient and low-carbon is a huge financial commitment for which the green recovery represents a once-in-a-generation opportunity. This implies digitization, and requires not only looking at large-scale renewable sources, which also remain relevant, but also at decentralized renewable sources, demand-side energy efficiency, and improved flexibility of power systems. In summary, efforts should be made to establish smart grids.[208] In this sense, digitization plays a fundamental role since it allows the creation of platforms for consumers to exchange electricity in the market. Additionally, it allows distribution companies to use tools such as machine learning and the internet of things to optimize the use and maintenance of the grid, as well as to optimize demand response.[209]

Fear of contagion threatens to undo much progress on moving individuals to public transport and reap the associated GHG emissions and urban air quality benefit. The resulting resurgence of private **urban mobility** should be catalyzed by a shift toward accessibility-based systems rather than an emphasis on the acquisition of private electric vehicles (footnote 209). This means moving beyond a product-based sector to a service-based sector. The progress of these measures will take place in interdependence with the progress of digitization, given the need for decentralized mobility exchange platforms.

[205]   IMF Fiscal Affairs. n.d. *Greening the Recovery*. https://www.imf.org/~/media/Files/Publications/covid19-special-notes/en-special-series-on-covid-19-greening-the-recovery.ashx?la=en.
[206]   ADB. 2020. *Green Finance Strategies for Post-COVID-19 Economic Recovery in Southeast Asia: Greening Recoveries for People and Planet*. https://www.adb.org/sites/default/files/publication/639141/green-finance-post-covid-19-southeast-asia.pdf.
[207]   OECD. 2020. *COVID-19 and the low-carbon transition: Impacts and possible policy responses*. http://www.oecd.org/coronavirus/policy-responses/covid-19-and-the-low-carbon-transition-impacts-and-possible-policy-responses-749738fc/#section-d1e539.
[208]   OECD. 2020. *Building back better: A sustainable, resilient recovery after COVID-19*. http://www.oecd.org/coronavirus/policy-responses/building-back-better-a-sustainable-resilient-recovery-after-covid-19-52b869f5/#section-d1e883.
[209]   T. Serebrisky. 2020. *Sustainable and digital infrastructure for the post-COVID-19 economic recovery of Latin America and the Caribbean: a roadmap to more jobs, integration and growth*. https://publications.iadb.org/publications/english/document/Sustainable-and-Digital-Infrastructure-for-the-Post-COVID-19-Economic-Recovery-of-Latin-America-and-the-Caribbean-A-Roadmap-to-More-Jobs-Integration-and-Growth.pdf.

6 Conclusion

ADB's Strategy 2030 sets seven operational priorities, each having its own operational plan. The operational plans contribute to ADB's vision to achieve prosperity, inclusion, resilience, and sustainability, and are closely aligned with Strategy 2030 principles and approaches (footnote 2). Tackling climate change, building climate and disaster resilience, and enhancing environmental sustainability are critical to achieving its Strategy 2030 vision of a prosperous, inclusive, resilient, and sustainable Asia and the Pacific (footnote 2). ADB has accordingly formulated three key strategic priorities under the third Operational Priority (OP3) of its Strategy 2030: (i) tackling climate change, (ii) building climate and disaster resilience, and (iii) enhancing environmental sustainability.

Digital technologies (DTs) have been hailed by many for their potential in being a catalyst to support climate action, build climate and disaster resilience, and enhance environmental sustainability at a reasonable cost. They can also provide economic and social progress and help achieve sustainable development targets.

The current landscape of DTs is quite wide and growing and can be broadly divided into three distinct stages. DTs of Stage I are currently being used for a range of supporting actions relevant to CCM, CCA, DRM, and environmental sustainability. These include the internet, satellite imagery, geographic information system (GIS), remote sensing, mobile phones, and databases. Stage II DTs are available but are not being utilized to their full potential to address CCM, CCA, DRM, and environmental sustainability, and include and include technologies such as social media, applications, and cloud computing. Finally, technologies of Stage III are expected to reach their commercial breakthroughs in the future and have the potential to significantly accelerate actions to address climate change mitigation and adaptation, environmental sustainability, and DRM. Stage III cutting-edge technologies include artificial intelligence, machine learning, IoT and smart systems; distributed ledger technologies and blockchain; big data and predictive analytics; virtual and mixed reality; and robotics and unmanned vehicles.

Evidently, and as expanded upon in the Appendix, ADB has a long experience with DTs from various stages. From 2010 to 2019, ADB supported 371 projects that included digital components (including 27 nonsovereign projects). Examples in the space of supporting ADB's OP3 include smart grid systems for renewable energy, smart sensors for nonrevenue water reductions, and real-time traffic control using intelligent transportation systems. ADB has launched Spatial Data Analysis Explorer (SPADE), a web-based platform to aid decision-making which can be utilized for consultation, project preparation, production of maps, and analysis of climate change impacts.[210] ADB is also applying Earth observation technologies and GISs in its projects, supported by partnerships with the European Space Agency and the Japan Aerospace Exploration Agency.[211] There is also a growing momentum on the use of DTs to enhance sustainable development and climate action among ADB DMCs as shown in this publication.

---

[210] ADB. Spatial Data Analysis Explorer. https://adb-spade.org/account/login/?next=/.
[211] ADB. 2019. *ADB Establishes High-Level Advisory Group for Digital Technology.* https://www.adb.org/news/adb-establishes-high-level-advisory-group-digital-technology.

This publication highlights the broad landscape of DTs countries can utilize to enhance the effectiveness of their efforts toward achieving OP3. This holistic view of DTs, along with the necessary considerations in selecting DTs to achieve development objectives, presented in this publication helps understand relative merits, costs, and limitations to the application of DTs such that decision makers can assess and evaluate the usefulness of DTs in their country and/or company context. For example, knowledge and analysis of key concepts such as Marginal Abatement Cost Curves and Marginal Adaptation Cost Curves can form the basis of related "appropriate technology" discussions for developing countries. DTs from difference stages can also be combined to achieve amplified impacts. DTs can also strengthen the ongoing efforts toward a green recovery and their impact can be maximized by leveraging public–private partnerships.

Despite the global recognition of the benefits of DT, however, the risks of digital divides suggest more investment is needed in policy and regulatory frameworks, infrastructure (including connectivity), applications (including contents and services), and capacities and skills (footnote 1). This means the successful application of these DTs requires an enabling environment that helps to overcome barriers which can range from lack of basic requirements such as national policies on DTs, lack of access to technology, or intermittency in nationwide internet coverage. Resolving these barriers requires specific strategies including, at the most, investments in basic DT infrastructure and education. In applying these DTs into developing country contexts, there is also a need for careful attention on issues related to capacity and usability as well as environmental, security, and privacy concerns.

Nonetheless, the strategic use of DT through well-designed initiatives can facilitate effective and efficient delivery of public services in the key sectors (footnote 1). It is in this context that ADB can support its DMCs in using DTs in a way that maximizes benefit to specific stakeholder groups while increasing the absorption capacity within countries. In doing so, DTs can effectively support ADB's strategic priorities under OP3 of ADB's Strategy 2030 which focuses on tackling climate change, building climate and disaster resilience, and enhancing environmental sustainability.

ADB Initiatives on Incorporating Digital Technology into Its Projects and Programs

This section provides a preliminary analysis on the integration of on the integration of digital technology (DT) into the operations and workstream of the Asian Development Bank (ADB). In 2003, ADB published its DT strategy, "Toward E-Development in Asia and the Pacific: A Strategic Approach for Information and Communication Technology." The strategy's three main goals are to

(i)   create an enabling environment through policy improvements, public institution strengthening, and relevant infrastructure provision;
(ii) build human resources for DT literacy and professional skills; and
(iii) develop DT applications and information content through ADB-supported projects and activities.

In 2016, the background report, Road to 2030: Information and Communications Technology in ADB's Corporate Strategy and Operations was released. This document highlighted the current state of DT, trends and applications, and recent and upcoming challenges within the sector.

On the basis of these strategies as well as the inclusion of DTs in high-level documents such as Strategy 2030, over the past 2 decades, ADB has extended DT-related loans, grants, and technical assistance projects to help develop and maintain sectors including DT infrastructure (e.g., telecommunications networks, mobile and wireless networks, broadband cable networks, satellite-based internet, data centers, last-mile internet connectivity, etc.); DT industries (e.g., DT centers of excellence, research/computer laboratories, DT-enabled industries such as business process outsourcing, knowledge process outsourcing, software parks, DT incubators, etc.); DT-enabled services (e.g., DT applications for agriculture, education, energy, finance, governance, health, transport, water and urban development, etc.); and DT policy, strategy, and capacity development (e.g., DT policy and strategy, telecommunications policy reform, universal access and service, DT road maps [national and local], DT regulations and laws, DT skills training and capacity building, etc.). From 2010 to 2019, ADB supported 371 projects that included digital components (including 27 nonsovereign projects) and their project type is visualized in Figure A1.

This is also true in the climate space. ADB has made efforts to diffuse the potential of new advanced technologies to support climate change adaptation and mitigation, focusing on the use of these technologies in disaster risk management (DRM) and preparedness, the adaptation of agricultural processes to new climate conditions, and the restructuring of the energy landscape through the implementation of smart cities. ADB has also incorporated innovative strategies for the application of DTs in its policies for key sectors which is visualized in Figure A2.

One notable initiative with regard to integration of DTs in decision-making is ADB's Spatial Data Analysis Explorer (SPADE), a web-based platform on a centralized cloud-based server that contains various geospatial data which can be utilized for consultation, project preparation, production of maps, and analysis of climate change impacts

Figure A1: ADB's Digital Technology Projects at a Glance
(2010–2019)

- ADB supported **371 DT projects** (including 27 nonsovereign projects)
  - ✓ 125 loans/grants
  - ✓ 246 TA projects
- **1 out of 10** (371 out of 3,709 ADB projects) had DT components.

	Typical DT interventions
Loan/Grant (nonsovereign)	• DT application, MIS development • Telecommunications
TA	• Pre-feasibility study • Advisory reports on policies and regulations • Training, knowledge sharing

DT Projects by Type

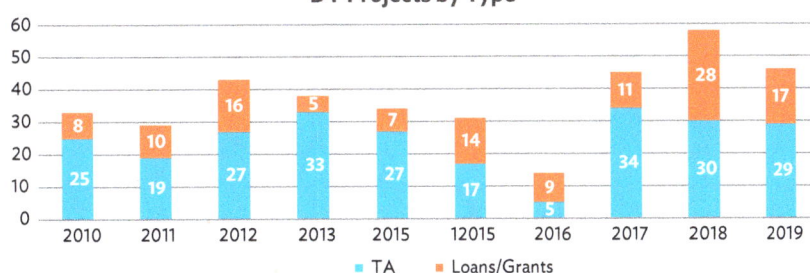

Year	TA	Loans/Grants
2010	25	8
2011	19	10
2012	27	16
2013	33	5
2015	27	7
12015	17	14
2016	5	9
2017	34	11
2018	30	28
2019	29	17

ADB = Asian Development Bank, DT = digital technology, MIS = management information system, TA = technical assistance.
Source: Asian Development Bank.

Figure A2: Digital Technology Projects per Sector (Loans and Grants) and Example Projects (2010–2019)

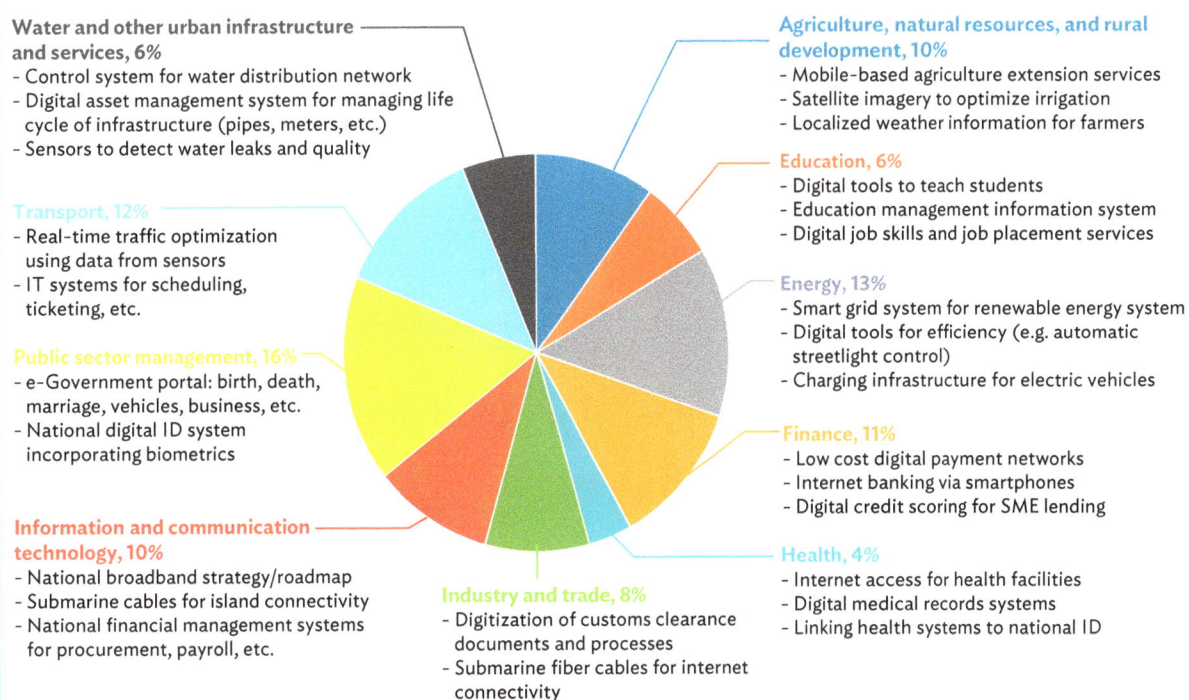

Water and other urban infrastructure and services, 6%
- Control system for water distribution network
- Digital asset management system for managing life cycle of infrastructure (pipes, meters, etc.)
- Sensors to detect water leaks and quality

Transport, 12%
- Real-time traffic optimization using data from sensors
- IT systems for scheduling, ticketing, etc.

Public sector management, 16%
- e-Government portal: birth, death, marriage, vehicles, business, etc.
- National digital ID system incorporating biometrics

Information and communication technology, 10%
- National broadband strategy/roadmap
- Submarine cables for island connectivity
- National financial management systems for procurement, payroll, etc.

Industry and trade, 8%
- Digitization of customs clearance documents and processes
- Submarine fiber cables for internet connectivity

Agriculture, natural resources, and rural development, 10%
- Mobile-based agriculture extension services
- Satellite imagery to optimize irrigation
- Localized weather information for farmers

Education, 6%
- Digital tools to teach students
- Education management information system
- Digital job skills and job placement services

Energy, 13%
- Smart grid system for renewable energy system
- Digital tools for efficiency (e.g. automatic streetlight control)
- Charging infrastructure for electric vehicles

Finance, 11%
- Low cost digital payment networks
- Internet banking via smartphones
- Digital credit scoring for SME lending

Health, 4%
- Internet access for health facilities
- Digital medical records systems
- Linking health systems to national ID

ID = identification, IT = information technology, SMEs = small and medium-sized enterprises.
Source: Asian Development Bank.

(footnote 211). Data is gathered from multiple sources including Earth observation, meteorological satellite data, and hydrometeorological data; climate change simulations from University of Tokyo's Data Integration and Analysis System (DIAS); socioeconomic data from field surveys; and a mobile application for data collection created by the Asian Institute of Technology (AIT). Upon gathering of data, it is integrated, standardized, and stored in a cloud-based system which enables queries, analysis, and visualization of impacts. Data is then presented as a website that is user-friendly and accessible on computers and mobile phones. Funded by the Urban Climate Change Resilience Trust Fund, SPADE can support ADB strategies and projects and scale up climate resilience thinking in urban planning. It is currently being piloted in five cities in two countries: Bagerhat and Patuakhali (Bangladesh) and Ha Giang, Hue, and Vinh Yen (Viet Nam).

ADB also promotes and encourages the implementation of projects with development organizations and other regional partners whose expertise and resources should be maximized in developing new ideas, innovative strategies, and technological platforms. ADB's assistance in DT for development is also carried out through partnership with the private sector and civil society organizations. ADB also encourages public–private partnerships to help promote investments in innovative technologies and strategies. Some notable partnerships and the work stream have been highlighted in the table below:

Table A1: ADB's Notable Partnerships in Advancing Digital Technology

Partner Organization	Joint Initiative
Asian and Pacific Centre for Information and Communication Technology (APCICT) National Information Society Agency (NIA)	• Improving Public Services through Information and Communication Technology • This project, aims to improve the capacity of governments in select DMCs in using ICTs for the better delivery of public services in key sector areas such as administration, education, health, transport, and agriculture.
International Telecommunication Union (ITU)	• Rural Information and Communication Technology Policy Advocacy, Knowledge Sharing, and Capacity Building • This technical assistance aims to contribute to improving policy, legal, and regulatory environments to make them more conducive to the rapid deployment of ICT infrastructure and services for rural development.
Japan Aerospace Exploration Agency (JAXA) European Space Agency (ESA)	• The initiative aims to increase cooperation with organizations that have space technology capability and use space-gathered information in ADB projects to improve planning, implementation, and monitoring. • Collaborations have helped in the application of satellite rainfall data to support the national flood forecasting agencies in Bangladesh, the Philippines, and Viet Nam; pilot testing of innovative new crop insurance products to give income protection from increasingly severe storms and natural disasters to small holder farmers in Bangladesh; and use freely available satellite data to develop regional food security cooperation in the six countries of the Greater Mekong Subregion, to name a few.
United Nations Economic and Social Commission for Asia and the Pacific (UNESCAP)	• JFICT 9068 Empowering the Rural Areas through Community e-Centers under the SASEC Program • This project aims to reduce poverty, empower communities, and improve the quality of life in rural areas in Bangladesh, Bhutan, India, and Nepal through increased ICT connectivity and accessibility.

ADB = Asian Development Bank, DMC = developing member country, ICT = information and communication technology, JFICT = Japan Fund for Information and Communication Technology, SASEC = South Asia Subregional Economic Cooperation.
Source: ADB. Digital technology: Policies and Strategies. https://www.adb.org/what-we-do/sectors/dt/strategy.

ADB has also initiated a number of technical assistance (TA) funds to help build the necessary capacity and technical knowledge in this area. For example, the Republic of Korea e-Asia and Knowledge Partnership Fund (EAKPF) was established in 2006 with the aim to bridge the digital divide and promote improved access to information and creating and sharing of knowledge through information and communication technology (ICT) in the Asia and Pacific region.[1] Similarly, the High-Level Technology (HLT) Fund is a multi-donor trust fund established in April 2017 which provides grant financing to promote the integration of HLT and innovative solutions into ADB, financed and administered sovereign and nonsovereign projects throughout the project cycle—from identification to implementation and operation.[2] More recently in 2019, ADB also initiated the ASEAN Australia Smart Cities Trust Fund (AASCTF) which supports activities that will enable cities to facilitate adaptation and adoption of digital solutions, systems and governance systems in the participating cities.[3]

Finally, a comprehensive, yet not exhaustive, list of institutional documents referring to ADB's strategic efforts in using and integrating digital technologies into climate change, environmental sustainability, and DRM work has been provided in the table below.

Table A2: Institutional Documents Referring to ADB's Strategic Efforts in Using and Integrating Digital Technologies into Climate Change, Environmental Sustainability, and Disaster Risk Management Work

Document Title	Date	Key Points
Leveraging Technology and Innovation for Disaster Risk Management and Financing	December 2020	This joint report by ADB and the OECD examines how emerging technologies and innovation can improve the management of disaster and climate risks and the availability and affordability of financial protection tools. It proposes strategic approaches to leveraging and integrating these technologies and innovations into effective DRM and financing. It suggests that governments focus on resilient communications infrastructure, technical skills, access to data and analytical technologies, insurance regulations, and user awareness and trust.
Digital Solutions: IOT For Digital Solutions for Real-Time Flood Early Warning Systems	September 2020	This 2-page document discusses how digital solutions such as flood early warning systems have proven to be critical tools in mitigating flood impact in cities, with specific use cases in Indonesia and Australia.
Frontiers of Water-Related Disaster Management and the Way Forward[a]	August 2020	This policy brief outlines the current limitations of water-related DRM and suggests some possible ways forward. It highlights the need for additional focus on identifying best practices regarding the use of advanced technology to enable effective governance for DRM, such as the effective use of science and technology in data integration and communication for disaster preparedness.
Asia and the Pacific Renewable Energy Status Report[b]	June 2020	The report presents the current status of renewable energy by examining the political landscape, investment flows, and the ways in which renewable energy improves access to energy. One of the regional challenges for the renewable energy sector is system reliability and the changing role of utilities. In this context, new digital technologies are identified as key factors for the shift toward decentralized energy access through the introduction of smart grids and the associated development of smart cities.
Irrigation Systems for Climate Change Adaptation in Viet Nam[c]	May 2020	This publication provides an overview of the Water Efficiency Improvement in Drought-Affected Provinces (WEIDAP) project, which envisages the use of remote sensing technology to map water productivity in five of the country's drought-prone provinces.

continued on next page

[1]   ADB. Republic of Korea e-Asia and Knowledge Partnership Fund. https://www.adb.org/what-we-do/funds/e-asia-and-knowledge-partnership-fund.
[2]   ADB. High-Level Technology Fund. https://www.adb.org/what-we-do/funds/high-level-technology-fund.
[3]   ADB. ASEAN Australia Smart Cities Trust Fund. https://www.adb.org/what-we-do/funds/asean-australia-smart-cities-fund.

Table A2 *continued*

Document Title	Date	Key Points
Climate-Smart Practices for Intensive Rice-Based Systems in Bangladesh, Cambodia, and Nepal[d]	October 2019	This report presents the results of initiatives to promote climate-friendly agricultural practices and technologies in Bangladesh, Cambodia, and Nepal. Based on the results, the identification of appropriate technologies is presented as one of the main recommendations. The use of ICT in the provision of agrometeorological consulting and market information is presented as an available high technology, which, however, is currently still little used.
Distributed Ledger Technologies for Developing Asia[e]	December 2017	This paper takes a first pass at assessing areas of implementation for distributed ledger (or blockchain) technology in the context of development finance. It identifies five use cases, including digital identity, trade finance, project aid monitoring, smart energy, and supply chain management. A discussion of the main benefits, risks and implementation challenges suggests that experimentation with distributed ledger technology can produce immediate significant benefits in some areas, while others require further research and investment, as well as additional technical, infrastructural, or regulatory development. Development lenders can play a role in helping unleash these technologies' positive developmental impact throughout the Asian region.
Embedding Community-Based Flood Risk Management in Investment: A Part-to-Whole Approach[f]	December 2017	This publication presents case studies in the People's Republic of China and Indonesia to show why holistic community-based approaches to flood risk management need to be a key feature of investment programs. In its recommendations, it emphasizes the need to promote technology transfer in the context of mitigation and disaster preparedness. Among the relevant technologies proposed are communication tools to facilitate early warning, including applications for cell phones, remote sensing, internet connections, and others.
Climate Change Profile of Pakistan[g]	August 2017	The report provides a comprehensive overview of science and policy on climate change in Pakistan. Accordingly, based on the needs and requirements of Pakistan's agriculture and water sector, adaptation technologies are identified and explained. Among them, the implementation of early warning systems and disaster risk information shows a strong interdependence with ICT.
Earth Observation for a Transforming Asia and Pacific[h]	March 2017	This report reviews the results of the Earth Observation for a Transforming Asia and Pacific initiative in promoting and demonstrating satellite-based Earth observation to support ADBs' investments in their DMCs. 10 of the 12 projects presented demonstrate the important influence of satellite-based environmental information in combating climate change, enhancing DRM, and environmental sustainability, including digital elevation models, as one of the main tools and sources of information.
Green Growth Opportunities for Asia[i]	January 2017	This publication assesses the low-carbon economy in Asia: its current size and its future potential. It analyzes the potential of Asian economies to benefit from the development and export of low-carbon technologies. It presents smart grid technologies as a further opportunity for pursuing green growth.
Technologies to Support Climate Change Adaptation in Developing Asia[j]	November 2014	This report presents technologies that can be used in climate change adaptation. Among the technologies presented, light detection and ranging, improved extreme weather forecasting, and early warning systems are highlighted as important applications of digital technologies to address the problem.
Space Technology and Geographic Information Systems Applications in ADB Projects	November 2014	This report provides an overview of the space technology and GIS applications in ADB to date by introducing some of the past and ongoing ADB projects that have applied space technology and/or GIS. It also includes information about how the technologies were applied, the service providers, and the cost for the application, so that practitioners including staff of development organizations and government staff in DMCs can easily apply similar technologies to their projects and/or daily operations.
Low-Carbon Green Growth in Asia: Policies and Practices[k]	June 2013	This book provides an overview of low-carbon policy initiatives taken by Asian countries at the national, sector, and local levels, while assessing successes, pinpointing deficits, and exploring new opportunities. Regarding the implementation of digital technologies, it proposes the promotion of public–private portfolio of large-scale integrated smart city and smart grid demonstration projects among the regional level actions for accelerating low-carbon green growth.

continued on next page

Table A2 *continued*

Document Title	Date	Key Points
Green Cities[l]	November 2012	This book presents ADB's endeavor to develop longer-term engagements in focus urban regions. Smart technologies are presented as one of the key mechanisms in achieving green cities.
Climate Risk and Adaptation in the Electric Power Sector[m]	July 2012	This report is intended to highlight and raise awareness of the vulnerability of the energy sector to climate change. Among the adaptation measures, it proposes for energy transmission and distribution are improved system management through "smart grids" investments, as the use of smart meters and other DTs should enable better energy management by consumers and utilities.
The Pilot Asia-Pacific Climate Technology Network and Finance Center[n]	July 2012	The flyer summarizes the Pilot Asia-Pacific Climate Technology Network and Finance Center, emphasizing the role of technology transfer in helping to address climate change. Innovative digital technologies mentioned as important triggers of boosting resilience are flood management and early warning technologies.
Policies and Practices for Low-Carbon Green Growth in Asia – Highlights[o]	May 2012	This book reviews and assesses the low-carbon and green policies and practices of Asian countries. It presents technology as one of the key drivers for low-carbon green growth, including the use of modern technologies such as "smart" electricity grids as a leapfrogging mechanism.

ADB = Asian Development Bank, DMC = developing member country, DRM = disaster risk management, GIS = geographic information system, ICT = information and communication technology, OECD = Organisation for Economic Co-operation and Development.

[a]   ADB Institute. 2020. *Frontiers of Water-Related Disaster Management and the Way Forward.* https://www.adb.org/sites/default/files/publication/630266/adbi-pb2020-4.pdf.

[b]   REN21. 2019. *Asia and the Pacific Renewable Energy Status Report.* https://www.adb.org/sites/default/files/publication/611911/asia-pacific-renewable-energy-status.pdf.

[c]   ADB. 2020. *Irrigation Systems for Climate Change Adaptation in Viet Nam.* https://www.adb.org/sites/default/files/publication/603186/irrigation-climate-change-adaptation-viet-nam.pdf.

[d]   ADB and OECD. 2020. *Leveraging Technology and Innovation for Disaster Risk Management and Financing.* https://www.adb.org/publications/technology-innovation-disaster-risk-mgt-financing.

[e]   B. Ferrarini, J. Maupin and M. Hinojales. 2017. Distributed ledger technologies for developing Asia. *ADB Economics Working Paper Series.* No. 533. Manila: Asian Development Bank.

[f]   ADB. 2020. *Digital Solutions: IOT For Digital Solutions for Real-Time Flood Early Warning Systems.* ASEAN Australia Smart Cities Trust Fund. https://events.development.asia/system/files/materials/2020/09/202009-aasctf-digital-solutions-brief-iot-real-time-flood-early-warning-systems.pdf.

[g]   ADB and IRRI. 2019. *Climate-Smart Practices for Intensive Rice-Based Systems in Bangladesh, Cambodia, and Nepal.* https://www.adb.org/sites/default/files/publication/533186/climate-smart-rice-systems-ban-cam-nep.pdf.

[h]   R. Osti. 2017. *Embedding Community-Based Flood Risk Management in Investment; A Part-To-Whole Approach.* https://www.adb.org/sites/default/files/publication/391681/eawp-012.pdf.

[i]   Q. Chaudry. 2017. Climate Change Profile of Pakistan. https://www.adb.org/sites/default/files/publication/357876/climate-change-profile-pakistan.pdf.

[j]   European Space Agency and Asian Development Bank. 2017. *Earth Observation For a Transforming Asia and Pacific: A Portfolio of Twelve Earth Observation Projects Supporting Asian Development Bank Activities.* https://www.adb.org/sites/default/files/publication/231486/earth-observation-asia-pacific.pdf.

[k]   S. Fankhauser, A. Kazaglis, and S. Srivastav. 2017. *Green Growth Opportunities for Asia.* https://www.adb.org/sites/default/files/publication/224391/ewp-508.pdf.

[l]   ADB. 2014. *Technologies to Support Climate Change Adaptation in Developing Asia.* https://www.adb.org/sites/default/files/publication/149400/technologies-climate-change-adaptation.pdf.

[m]   ADB. 2014. *Space Technology and Geographic Information Systems Applications in ADB Projects.* https://www.adb.org/sites/default/files/publication/148901/space-technology-and-gis-applications-adb-projects.pdf.

[n]   ADB and ADBI. 2013. *Low-Carbon Green Growth in Asia: Policies and Practices.* https://www.adb.org/sites/default/files/publication/159319/adbi-low-carbon-green-growth-asia.pdf.

[o]   ADB. 2012. *Green Cities.* https://www.adb.org/sites/default/files/publication/30059/green-cities.pdf.

[p]   ADB. 2012. *Climate Risk and Adaptation in the Electric Power Sector.* https://www.adb.org/sites/default/files/publication/29889/climate-risks-adaptation-power-sector.pdf.

[q]   ADB, UNEP and GEF. 2012. *The Pilot Asia-Pacific Climate Technology Network and Finance Center.* https://www.adb.org/sites/default/files/publication/29975/pilot-asia-pacific-climate-technology-flyer.pdf.

[r]   ADB and ADBI. 2012. *Policies And Practices For Low- Carbon Green Growth ian Asia.* https://www.adb.org/sites/default/files/publication/159328/adbi-policies-practices-low-carbon-green-growth-asia-highlights.pdf.

Source: Author's compilation.

References

Acciona. 2020. *Top 10 Apps for Sustainable Living.* https://www.activesustainability.com/sustainable-life/top-10-apps-for-sustainable-living/. (accessed 20 October 2020).

F. Adamo et al. 2015. A Smart Sensor Network for Sea Water Quality Monitoring. *IEEE Sensors Journal.* Vol. 15 (5). pp. 2,514–2,522.

African Development Bank (AfDB). 2012. *Information Technology Strategy 2013–2015.* Abidjan. https://www.afdb.org/fileadmin/uploads/afdb/Documents/Policy-Documents/Information%20Technology%20Strategy%202013-2015%20-%20Revised.pdf.

————. 2015. *Senegal–Digital Technology Park Project.* Abidjan. https://projectsportal.afdb.org/dataportal/VProject/show/P-SN-G00-001.

————. 2017. *Tunisia–Support Project for the Implementation of the "Digital Tunisia 2020" National Strategic Plan.* Abidjan. https://projectsportal.afdb.org/dataportal/VProject/show/P-TN-G00-003.

————. 2019. *Central Africa Regional Integration Strategy Paper 2019–2025.* Abidjan. https://www.afdb.org/en/documents/central-africa-regional-integration-strategy-paper-2019-2025.

————. 2019. *Multinational–Project for Digitisation of Government Payments in the Mano River Union (DIGIGOV_MRU) Countries–Guinea.* Abidjan. https://projectsportal.afdb.org/dataportal/VProject/show/P-Z1-HB0-064.

————. 2020. *Africa Digital Financial Inclusion Facility (ADFI).* Abidjan. https://www.afdb.org/en/adfi.

————. 2020. *Information and communication technology.* Abidjan. https://www.afdb.org/en/topics-and-sectors/sectors/information-communication-technology.

————. 2020. *Kenya–Rural Transformation Centres Digital Platform Project.* Abidjan. https://projectsportal.afdb.org/dataportal/VProject/show/P-KE-AA0-022.

Asian Development Bank. 2012. *Climate Risk and Adaptation in the Electric Power Sector.* https://www.adb.org/sites/default/files/publication/29889/climate-risks-adaptation-power-sector.pdf.

————. 2012. *Green Cities.* https://www.adb.org/sites/default/files/publication/30059/green-cities.pdf.

————. 2012. *The Pilot Asia-Pacific Climate Technology Network and Finance Center.* https://www.adb.org/sites/default/files/publication/29975/pilot-asia-pacific-climate-technology-flyer.pdf.

———. 2013. *Low-Carbon Green Growth in Asia: Policies and Practices.* https://www.adb.org/sites/default/files/publication/159319/adbi-low-carbon-green-growth-asia.pdf.

———. 2014. *Space Technology and Geographic Information Systems Applications in ADB Projects.* https://www.adb.org/sites/default/files/publication/148901/space-technology-and-gis-applications-adb-projects.pdf.

———. 2014. *Technologies to Support Climate Change Adaptation in Developing Asia.* https://www.adb.org/sites/default/files/publication/149400/technologies-climate-change-adaptation.pdf.

———. 2016. *Earth Observation for a Transforming Asia and Pacific.* https://www.adb.org/sites/default/files/publication/231486/earth-observation-asia-pacific.pdf.

———. 2017. *Climate Change Profile of Pakistan.* https://www.adb.org/sites/default/files/publication/357876/climate-change-profile-pakistan.pdf.

———. 2017. *Embedding Community-Based Flood Risk Management in Investment: A Part-to-Whole Approach.* https://www.adb.org/sites/default/files/publication/391681/eawp-012.pdf.

———. 2017. *Green Growth Opportunities for Asia.* https://www.adb.org/sites/default/files/publication/224391/ewp-508.pdf.

———. 2018. *Strategy 2030: Solomon Islands, Timor-Leste, Tonga, and Vanuatu.* Manila. https://www.adb.org/about/strategy-2030-operational-priorities.

———. 2018. *Digital Agenda 2030: Special Capital Expenditure Requirements for 2019–2023.* https://www.adb.org/sites/default/files/institutional-document/472371/adb-digital-agenda-2030.pdf.

———. 2018. *Strategy 2030: Achieving a Prosperous, Inclusive, Resilient, and Sustainable Asia and the Pacific.* https://www.adb.org/sites/default/files/institutional-document/435391/strategy-2030-main-document.pdf.

———. 2019. ADB Establishes High-Level Advisory Group for Digital Technology. 2 September. https://www.adb.org/news/adb-establishes-high-level-advisory-group-digital-technology.

———. 2019. *Asia and the Pacific: Renewable Energy Status Report.* https://www.adb.org/sites/default/files/publication/611911/asia-pacific-renewable-energy-status.pdf.

———. 2019. *Climate-Smart Practices for Intensive Rice-Based Systems in Bangladesh, Cambodia, and Nepal.* https://www.adb.org/sites/default/files/publication/533186/climate-smart-rice-systems-ban-cam-nep.pdf.

———. 2019. *Strategy 2030: Operational Plan for Priority 3.* Manila. https://www.adb.org/sites/default/files/institutional-document/495961/strategy-2030-op3-climate-change-resilience-sustainability.pdf.

———. 2020. *ADB's Focus on Digital Technology.* https://www.adb.org/what-we-do/sectors/dt/main.

———. 2020. *Digital Solutions: IOT For Digital Solutions for Real-Time Flood Early Warning Systems.* ASEAN Australia Smart Cities Trust Fund https://events.development.asia/system/files/materials/2020/09/202009-aasctf-digital-solutions-brief-iot-real-time-flood-early-warning-systems.pdf.

———. 2020. *Frontiers of Water-Related Disaster Management and the Way Forward.* https://www.adb.org/sites/default/files/publication/630266/adbi-pb2020-4.pdf.

———. 2020. *Green Finance Strategies for Post-Covid-19 Economic Recovery in Southeast Asia*. https://www.adb.org/sites/default/files/publication/639141/green-finance-post-covid-19-southeast-asia.pdf.

———. 2020. *Irrigation Systems for Climate Change Adaptation in Viet Nam*. https://www.adb.org/sites/default/files/publication/603186/irrigation-climate-change-adaptation-viet-nam.pdf.

ADB. Spatial Data Analysis Explorer. https://adb-spade.org/account/login/?next=/.

ADB and IADB. 2012. *Policies and Practices for Low-Carbon Green Growth in Asia*. https://www.adb.org/sites/default/files/publication/159328/adbi-policies-practices-low-carbon-green-growth-asia-highlights.pdf.

ADB and OECD. 2020. *Leveraging Technology and Innovation for Disaster Risk Management and Financing*. https://www.adb.org/publications/technology-innovation-disaster-risk-mgt-financing.

M. Agarwal. 2018. *Blockchain: India likely to see brain drain as 80% developers may move abroad. Inc42*. https://inc42.com/buzz/blockchain-india-likely-to-suffer-brain-drain-as-80-developers-prepare-to-move-abroad/.

Air-quality.com. 2020. https://air-quality.com/ (accessed 20 October 2020).

A.R. Al-Ali et al. 2017. A smart home energy management system using IoT and big data analytics approach. *IEEE Transactions on Consumer Electronics*. November. Vol. 63 (4). pp. 426–434.

K. Albus, R. Thompson, and F. Mitchell. 2019. Usability of Existing Volunteer Water Monitoring Data: What Can the Literature Tell Us? *Citizen Science: Theory and Practice*. Vol 4 (1). https://theoryandpractice.citizenscienceassociation.org/articles/10.5334/cstp.222/.

AllAfrica. 2015. *Africa: Chinese Internet Giant Tencent to Promote Wildlife Conservation in Kenya*. 16 December. http://allafrica.com/stories/201512160323.html (accessed 10 October 2020).

Amnesty International. 2020. DRC: Alarming research shows long lasting harm from cobalt mine abuses. 6 May. https://www.amnesty.org/en/latest/news/2020/05/drc-alarming-research-harm-from-cobalt-mine-abuses/.

W. Ashford. 2019. Social media and enterprise apps pose big security risks. *ComputerWeekly.com*. https://www.computerweekly.com/news/252469873/Social-media-and-enterprise-apps-pose-big-security-risks.

AWS. 2020. *What is Cloud Computing?* https://aws.amazon.com/de/what-is-cloud-computing/ (accessed 22 October 2020).

K. Bayne et al. 2012. The introduction of robotics for New Zealand forestry operations: Forest sector employee perceptions and implications. *Technology in Society*. Vol. 34 (2).

Y. Bhatia. 2018. Sudden escalation of the feature phone market in India. *Telecom Economic Times*. 9 April. https://telecom.economictimes.indiatimes.com/tele-talk/sudden-escalation-of-the-feature-phone-market-in-india/2984.

BBVA. 2019. *The Internet of things and its impact on sustainability*. 21 November. https://www.bbva.com/en/the-internet-of-things-and-its-impact-on-sustainability/.

C. Beam. 2019. Soon satellites will be able to watch you everywhere all the time. *MIT Review*. 26 June. https://www.technologyreview.com/2019/06/26/102931/satellites-threaten-privacy/.

L. Bellatreche, P. Valduriez and T. Morzy. 2018. Advances in Databases and Information Systems. *Information System Frontiers*. Vol. 20, pp. 1–6.

BlockchainHub Berlin. 2018. *Blockchain & Sustainability*. 11 August. https://blockchainhub.net/blog/blog/blockchain-sustainability-programming-a-sustainable-world/.

C. Blundell, K.-T. Lee and S. Nykvist. 2016. Digital learning in schools: Conceptualizing the challenges and influences on teacher practice. *Journal of Information Technology Education: Research.* Vol. 15, pp. 535–560.

J.D. Borrero and A. Zabalo. 2020. An autonomous wireless device for real-time monitoring of water needs. Sensors, 20(7), p. 2078.

T. Bradley. 2020. Cybersecurity Priorities are A Matter of Perspective. *Forbes.* https://www.forbes.com/sites/tonybradley/2020/02/05/cybersecurity-priorities-are-a-matter-of-perspective/#5575d6345d17.

T. Brady. 2019. How emerging tech can counteract climate change. *GreenBIZ.* 23 January. https://www.greenbiz.com/article/how-emerging-tech-can-counteract-climate-change.

B. Brauer et al. 2016. Green by App: The Contribution of Mobile Applications to Environmental Sustainability. *Proceedings of 20th Pacific Asia Conference on Information Systems (PACIS July 2016),* Chiayi City. https://www.researchgate.net/publication/308708034_GREEN_BY_APP_THE_CONTRIBUTION_OF_MOBILE_APPLICATIONS_TO_ENVIRONMENTAL_SUSTAINABILITY.

Breeze Technologies. 2020. https://www.breeze-technologies.de/solutions/urban-air-quality/ (accessed 22 October 2020).

G. Brodnig et al. 2017. Bridging the Gap: The Role of Spatial Information Technologies in the Integration of Traditional Environmental Knowledge and Western Science. *Information Systems in Developing Countries.* Vol. 1 (1).

E. Bruzelius et al. 2019. Satellite images and machine learning can identify remote communities to facilitate access to health services. *Journal of the American Medical Informatics Association.* August/September. Vol. 26 (8–9), pp. 806–812. https://doi.org/10.1093/jamia/ocz111.

K. Buchholz. 2020. Where 5G Technology Has Been Deployed. *Statista.* 14 October. https://www.statista.com/chart/23194/5g-networks-deployment-world-map.

B. Buntz. 2019. How IoT Technology Can Help the Environment. *IoTWorldToday.* 19 December. https://www.iotworldtoday.com/2019/12/19/how-iot-technology-can-help-the-environment/.

Business Insider. 2020. *The security and privacy issues that come with the Internet of Things.* https://www.businessinsider.com/iot-security-privacy?r=DE&IR=T.

———. 2020. Ocean Cleanup Project. *In one summer, they were able to label more than 30,000 photos.* https://www.businessinsider.com/microsoft-machine-learning-for-the-ocean-cleanup-project-2020-10?r=DE&IR=T#in-one-summer-they-were-able-to-label-more-than-30000-photos-20.

Business Today India. 2019. *80% of Indian engineers not fit for jobs, says survey.* 25 March. https://www.businessinsider.com/microsoft-machine-learning-for-the-ocean-cleanup-project-2020-10?r=DE&IR=T#in-one-summer-they-were-able-to-label-more-than-30000-photos-20.

P. Campbell. 2020. EV supply chains seek clearer visibility with blockchain. *Financial Times*. https://www.ft.com/content/3652b68e-206f-4e3a-9b3a-4d5ec9a285b7.

Capture. 2020. *Planet-friendly living, made possible.* https://www.thecapture.club/.

CGTN. 2019. *Feeding 1.4 Billion: Smart farming in China's big grain warehouse.* https://news.cgtn.com/news/2019-12-13/Feeding-1-4-Billion-Smart-farming-in-China-s-big-grain-warehouse-MohBFcaajK/index.html.

CGIAR. PlantVillage Nuru: AI for pest & disease monitoring. https://bigdata.cgiar.org/inspire/inspire-challenge-2017/pest-and-disease-monitoring-by-using-artificial-intelligence (accessed 18 April 2021)

J. Charness. 2019. How Oracle and The World Bee Project are Using AI to Save Bees. *Oracle AI and Data Science Blog*. https://blogs.oracle.com/datascience/how-oracle-and-the-world-bee-project-are-using-ai-to-save-bees-v2.

R. Cho. 2018. Artificial Intelligence–A Game Changer for Climate Change and the Environment. *Earth Institute, Columbia University*. 5 June. https://blogs.ei.columbia.edu/2018/06/05/artificial-intelligence-climate-environment/.

J. Clement. 2020. Number of worldwide social media users 2020, by region. *Statista*. 15 July. https://www.statista.com/statistics/454772/number-social-media-user-worldwide-region/.

———. 2020. Worldwide digital population as of July 2020. *Statista*. 24 July. https://www.statista.com/statistics/617136/digital-population-worldwide/.

Green for Growth Fund. Clim@2020 Finalist: Forest Guard. https://www.ggf.lu/press/news/news-detail/clim-competition-2020-announces-15-finalists-with-circular-economy-solutions-the-most-popular-theme-30.

N. Coca. 2015. *Palm Oil from freshly-burned Land: Coming to a Grocery Store near you*. 13 November. http://www.triplepundit.com/2015/11/palm-oil-freshly-burned-land-coming-grocery-store-near/ (accessed 11 October 2020).

U.W. Cohan. 2017. Assessing the Differences in Bitcoin & Other Cryptocurrency Legality Across National Jurisdictions. *SSRN* (revised 2 April 2020). https://papers.ssrn.com/sol3/papers.cfm?abstract_id=3042248.

P. Collela. 2017. *5G and IoT: Ushering in a new era*. 30 March. https://www.ericsson.com/en/about-us/company-facts/ericsson-worldwide/india/authored-articles/5g-and-iot-ushering-in-a-new-era.

B.D. Collins and R.W. Jibson. 2015. Assessment of existing and potential landslide hazards resulting from the April 25, 2015, Gorkha, Nepal earthquake sequence. *US Geological Survey* (revised 24 August 2015). https://pubs.er.usgs.gov/publication/ofr20151142.

F. Dahlqvist et al. 2019. Growing opportunities in the Internet of Things. McKinsey & Company. https://www.mckinsey.com/industries/private-equity-and-principal-investors/our-insights/growing-opportunities-in-the-internet-of-things.

B. D'Amico et al. 2018. Machine Learning for Sustainable Structures: A Call for Data. *Research Gate*. November. https://www.researchgate.net/publication/329048005_Machine_Learning_for_Sustainable_Structures_A_Call_for_Data.

Danish Technological Institute. 2020. *Robots with Artificial Intelligence to Sort Hazardous Waste.* https://www.dti.dk/specialists/robots-with-artificial-intelligence-to-sort-hazardous-waste/38310 (accessed 10 October 2020).

Datta. 2019. What is the difference between Virtual Reality, Augmented Reality and Mixed Reality. GeoSpatial World. 05 September. https://www.geospatialworld.net/article/difference-virtual-reality-augmented-reality-mixed-reality/.

S. Davidson et al. 2016. Disrupting Governance: The New Institutional Economics of Distributed Ledger Technology. *SSRN.* 22 July. https://papers.ssrn.com/sol3/papers.cfm?abstract_id=2811995.

Department for Communities and Local Government. 2009. *Multi-criteria analysis: a manual.* London. http://eprints.lse.ac.uk/12761/1/Multi-criteria_Analysis.pdf.

District Farms. 2020. *Clean, Delicious, Healthy Leafy Green.* http://district.farm/.

J. Dubow. 2014. *The World Bank Big data and urban mobility*. Cairo. 2 June. https://www.worldbank.org/content/dam/Worldbank/Feature%20Story/mena/Egypt/Egypt-Doc/Big-Data-and-Urban-Mobility-v2.pdf.

DW.com. 2020. Indonesia: A smartphone app to manage household waste. https://www.dw.com/en/indonesia-a-smartphone-app-to-manage-household-waste/av-51558686 (accessed 12 October 2020).

EcoMENA. 2020. *The Role of IoT in Sustainable Development.* https://www.ecomena.org/internet-of-things/.

Economist. 2017. China may match or beat America in AI: Its deep pool of data may let it lead in artificial intelligence. 15 July. https://www.economist.com/news/business/21725018-its-deep-pool-data-may-let-it-lead-artificial-intelligence-china-may-match-or-beat-america.

L. Edwards. 2020. Blockchain can be a vital tool to boost sustainability. *Sustainability Times.* 14 April. https://www.sustainability-times.com/sustainable-business/blockchain-can-be-a-vital-tool-to-boost-sustainability/.

L. Einav and J. Levin. 2013. *The data revolution and economic analysis, NBER Innovation Policy and the Economy Conference.* April 2013. http://www.nber.org/papers/w19035.

S. Ejiaku. 2014. *Technology Adoption: Issues and Challenges in Information Technology Adoption in Emerging Economies.* https://core.ac.uk/download/pdf/55335431.pdf.

P. Ekins, F. Kesicki, and A.Z.P. Smith. 2011. Marginal Abatement Cost Curves: A Call for Caution. *UCL Energy Institute.* London. April. http://www.homepages.ucl.ac.uk/~ucft347/MACCCritGPUKFin.pdf.

Energizer. 2020. *KaiOS Operating System.* https://www.energizeyourdevice.com/en/mobiles/product/details/kaios-operating-system/.

V. Eory et al. 2017. Marginal abatement cost curves for agricultural climate policy. *State-of-the-art, lessons learnt, and future potential.* https://www.sciencedirect.com/science/article/abs/pii/S095965261830283X.

EOS Earth Observing System. 2019. Satellite Data: What Spatial Resolution is enough? 12 April. https://eos.com/blog/satellite-data-what-spatial-resolution-is-enough-for-you/.

S. Ermon. 2018. Machine Learning and Decision Making for Sustainability. *Stanford University. IJCAI.* 13 July. https://cs.stanford.edu/~ermon/slides/ermon_ijcai_early.pdf.

ETHW. 2017. *Millimeter Waves.* 12 April. https://ethw.org/Millimeter_Waves.

EU Commission Press. 2020. Taxation and the European Green Deal: Commission launches public consultations on Energy Taxation Directive and a Carbon Border Adjustment Mechanism, EU Commission Press. *PubAffairs Bruxelles.* 23 July. https://ec.europa.eu/taxation_customs/news/commission-launches-public-consultations-energy-taxation-and-carbon-border-adjustment-mechanism_en.

European Commission–Joint Research Centre–Institute for Environment and Sustainability. 2011. *ILCD Handbook: Recommendations for Life Cycle Impact Assessment in the European context.* Luxemburg. First edition November 2011. https://eplca.jrc.ec.europa.eu/uploads/ILCD-Recommendation-of-methods-for-LCIA-def.pdf.

European Commission. 2017. *Measuring the economic impact of cloud computing in Europe.* https://ec.europa.eu/digital-single-market/en/news/measuring-economic-impact-cloud-computing-europe.

———. 2019. *Cloud computing: A Different Way of Using IT.* 18 September. https://ec.europa.eu/digital-single-market/en/news/cloud-computing-brochure.

———. 2020. *Cloud Computing* (updated 15 October 2020). https://ec.europa.eu/digital-single-market/en/cloud.

Fashionomics Africa. 2020. https://fashionomicsafrica.org/ (accessed 10 October 2020).

Fast Company. *2014. RFID-Tagged Rhinos and Smart Watering Holes: The Google-Funded Tech Fighting Poaching.* 2 July. https://www.fastcompany.com/3026125/rfid-tagged-rhinos-and-smart-watering-holes-the-google-funded-tech-fighting-poaching.

Fastmetrics. 2020. *Average Internet Speeds by Country.* https://www.fastmetrics.com/internet-connection-speed-by-country.php. (accessed 15 October 2020).

D. Fazio. n.d. Green Crops. https://unsplash.com/photos/oK9EKfqv8HE.

Federal Office for Information Security. 2020. *What is Cloud Computing?* https://www.bsi.bund.de/EN/Topics/CloudComputing/Basics/Basics_node.html (accessed 22 October 2020).

B. Ferrarini, J. Maupin and M. Hinojales. 2017. Distributed ledger technologies for developing Asia. *ADB Economics Working Paper Series.* No. 533. Manila: Asian Development Bank.

G. Fischer et al. 2002. Climate Change and Agricultural Vulnerability. *IIASA.* Laxenburg (revised 10 May 2002). http://pure.iiasa.ac.at/id/eprint/6670/.

Food and Agriculture Organization (FAO). 2018. Water Productivity Through Open Access of Remotely Sensed Derived Data Portal (WaPOR). http://www.fao.org/3/CA1081EN/ca1081en.pdf.

Forest Guard. Smart Sensor Systems. 2020. https://forestguard228852108.wordpress.com/.

Gartner. 2019. Gartner 2019 Hype Cycle Shows Most Blockchain Technologies Are Still Five to 10 Years Away From Transformational Impact. *Stamford, Conn.* 8 October. https://www.gartner.com/en/newsroom/press-releases/2019-10-08-gartner-2019-hype-cycle-shows-most-blockchain-technologies-are-still-five-to-10-years-away-from-transformational-impact.

Gartner. 2013. *Big data.* http://www.gartner.com/it-glossary/big-data/.

D. Gayle. 2016. Smart devices 'too dumb' to fend off cyber-attacks say experts. *The Guardian.* 22 October. https://www.theguardian.com/technology/2016/oct/22/smart-devices-too-dumb-to-fend-off-cyber-attacks-say-experts.

GCF Global. 2020. *Computer Science-Algorithms.* https://edu.gcfglobal.org/en/computer-science/algorithms/1/.

———. 2020. *What is an Application?* https://edu.gcfglobal.org/en/computerbasics/understanding-applications/1/.

———. 2020. *What is the Internet?* https://edu.gcfglobal.org/en/internetbasics/what-is-the-internet/1/.

GeSI. 2020. *Digital Solutions for supporting the Achievement of the NDC.* https://gesi.org/research/digital-solutions-for-supporting-the-achievement-of-the-ndc.

N. Gibbs. 2020. Automakers call out industry's weak spots: Companies seek more ethical, resilient supply chains. *Automotive News. Detroit.* Vol. 94 (6947). 17 August. p. 8. https://www.autonews.com/suppliers/automakers-seek-more-ethical-resilient-supply-chains.

GIS Geography. 2020. *Sentinel 2 Bands and Combinations.* 18 October. https://gisgeography.com/sentinel-2-bands-combinations/.

Gill et al. 2020. The Economics of AI-Based Technologies: A Framework and an Application to Europe. https://papers.ssrn.com/sol3/papers.cfm?abstract_id=3660114#.

Government of Australia, Great Barrier Reef Marine Park Authority. 2020. *Eye on the reef.* http://www.gbrmpa.gov.au/our-work/eye-on-the-reef (accessed 5 October 2020).

Government of New Zealand, Ministry of Education. *Code Avengers.* https://www.codeavengers.com/profile#all (accessed 11 October 2020).

GreentownLabs. 2020. *Autonomous Marine Systems.* https://greentownlabs.com/members/autonomous-marine-systems/.

R. Gupta. 2019. The State of Artificial Intelligence Development in India. *ViaNews.* https://via.news/asia/artificial-intelligence-development-india/.

T. Haigh. 2011. The history of information technology. *Annual Review of Information Science and Technology.* Vol. 45 (1).

HaiThanh. 2008. Hanoi–after the rain. *Flickr.com.* 31 October. https://www.flickr.com/photos/52621576@N00/2988954866.

D.G. Harkut and K. Kasat. 2019. Introductory Chapter: Artificial Intelligence-Challenges and Applications. In Artificial Intelligence-Scope and Limitations. IntechOpen.

Harvest AI. 2020. *Next Generation Farming.* https://harvest-ai.com/.

HERE Mobility. 2020. Smart Traffic Systems 101: Components, Benefits, and the Big Data Connection. https://mobility.here.com/learn/smart-transportation/smart-traffic-systems-101-components-benefits-and-big-data-connection#pgid-1656.

S. Heron et al. 2016. Validation of Reef-Scale Thermal Stress Satellite Products for Coral Bleaching Monitoring. *Remote Sensing 2016*. Vol 8 (1). p. 59. https://earth.org/satellite-imagery-helping-to-detect-plastic-pollution-in-the-ocean/.

C. Hinga. 2020. *Saying good-bye to Crowdmap.com.* 30 September. https://www.ushahidi.com/.

A. Holst. 2020. AI Implementation in Organizations Worldwide. *Statista.* https://www.statista.com/statistics/1133015/statements-best-describes-ai-implementation-in-organizations/ (accessed 25 November 2020).

M. Hossain et al. 2017. How do digital platforms for ideas, technologies, and knowledge transfer act as enablers for digital transformation? *Technology Innovation Management Review.* September. Vol. 7 (9). pp. 55–60.

HOT. 2020. *Volunteer Mappers.* https://www.hotosm.org/volunteer-opportunities/volunteer-mappers/ (accessed 15 October).

X. Hou et al. 2019. SolarNet: A Deep Learning Framework to Map Solar Power Plants in China from Satellite Imagery. Version v2. *Cornell University* (revised 10 December 2019). https://arxiv.org/abs/1912.03685.

B.C. Howard. 2015. Vanuatu Puts Drones in the Sky to see Cyclone Damage. *National Geographic.* 8 April. https://www.nationalgeographic.com/news/2015/04/150406-vanuatu-cyclone-pam-relief-drones-uavs-crisis-mapping-patrick-meier/.

Inter-American Development Bank. 2013. *Development of a methodology for the construction of marginal abatement cost curves for GHG emissions, including the uncertainty linked to the main mitigation parameters.* Washington, DC. https://publications.iadb.org/publications/spanish/document/Desarrollo-de-una-metodolog%C3%ADa-para-la-construcci%C3%B3n-de-curvas-de-abatimiento-de-emisiones-de-GEI-incorporando-la-incertidumbre-asociada-a-las-principales-variables-de-mitigaci%C3%B3n.pdf.

———. 2017. *Digital Finance. New Times, New Challenges, New Opportunities.* Washington, DC. https://publications.iadb.org/publications/english/document/Digital-Finance-New-Times-New-Challenges-New-Opportunities.pdf.

———. 2017. *How Information and Communications Technology (ICT) is Poised to Transform the Delivery of Energy Services.* Washington, DC. https://publications.iadb.org/en/how-information-and-communications-technology-ict-poised-transform-delivery-energy-services.

———. 2017. *Innovation, Science, and Technology Sector Framework Document.* Washington, DC. https://www.iadb.org/en/sector/science-and-technology/sector-framework.

———. 2018. *Technology for Climate Action in Latin America and the Caribbean: How ICT and Mobile Solutions Contribute to a Sustainable, Low-Carbon Future.* Washington, DC. https://publications.iadb.org/en/technology-climate-action-latin-america-and-caribbean-how-ict-and-mobile-solutions-contribute.

———. 2019. *Approach to Digital Transformation: Guidelines and Recommendations.* Washington, DC. https://publications.iadb.org/publications/english/document/Approach_to_Digital_Transformation_Guidelines_and_Recommendations.pdf.

———. 2019. *Guiding the digital transformation of Education Management and Information Systems (SIGEDs).* Washington, DC. https://publications.iadb.org/publications/english/document/From_Paper_to_the_Cloud_Guiding_the_Digital_Transformation_of_Education_Management_and_Information_Systems_SIGEDs.pdf.

———. 2019. *Quantum Technologies: Digital Transformation, Social Impact, and Cross-Sector Disruption.* Washington, DC. https://publications.iadb.org/en/quantum-technologies-digital-transformation-social-impact-and-cross-sector-disruption.

———. 2019. *Second Update to the Institutional Strategy: Summary.* Washington, DC. https://publications.iadb.org/en/second-update-institutional-strategy-summary.

———. 2019. *The Future We Aim for: Towards a "360 Resilience" Development Paradigm for the Caribbean.* Washington, DC. https://publications.iadb.org/en/future-we-aim-towards-360-resilience-development-paradigm-caribbean.

———. 2020. *Digital economy and technology in the service of the region's development. Economic report on Central America and the Dominican Republic.* Washington, DC. https://publications.iadb.org/publications/english/document/Digital_Economy_and_Technology_in_The_Service_of_The_Regions_Development_Economic_Report_on_Central_America_and_the_Dominican_Republic.pdf.

———. 2020. *New Technologies and Trade: New Determinants, Modalities, and Varieties.* Washington, DC. https://publications.iadb.org/publications/english/document/New_Technologies_and_Trade_New_Determinants_Modalities_and_Varieties_en.pdf.

———. 2020. *Sustainable and digital infrastructure for the post-COVID-19 economic recovery of Latin America and the Caribbean: a roadmap to more jobs, integration and growth.* Washington, DC. https://publications.iadb.org/publications/english/document/Sustainable-and-Digital-Infrastructure-for-the-Post-COVID-19-Economic-Recovery-of-Latin-America-and-the-Caribbean-A-Roadmap-to-More-Jobs-Integration-and-Growth.pdf.

R.T. Ilieva. 2018. Social Media Data for Urban Sustainability. *Springer Nature Sustainability Community*. 15 October. https://sustainabilitycommunity.springernature.com/posts/39900-social-media-data-for-urban-sustainability.

IMF. Fiscal Affairs Department. 2020. *Special Series on Fiscal Policies to Respond to COVID-19. Greening the Recovery.* https://www.imf.org/~/media/Files/Publications/covid19-special-notes/en-special-series-on-covid-19-greening-the-recovery.ashx?la=en.

S. Imran et al. 2020. Frontier technologies to protect the environment and tackle climate change. *ITU*. April. Geneva. https://www.itu.int/en/action/environment-and-climate-change/Documents/frontier-technologies-to-protect-the-environment-and-tackle-climate-change.pdf.

Infosecurity. 2018. How to Balance Security with Digital Transformation. https://www.infosecurity-magazine.com/opinions/balance-security-digital/.

Internetworldstats.com. 2020. *Low levels of internet penetration persist e.g. in Afghanistan, Cambodia, or Pakistan.* https://www.internetworldstats.com/asia.htm.

Investopedia. n.d. *Social Media.* https://www.investopedia.com/terms/s/social-media.asp (accessed 18 April 2021)

IPE. 2020. *Weather-Air-Blue Map.* http://wwwen.ipe.org.cn/appdownload30_en/pc/index.html (accessed 5 October 2020).

IQAir. 2020. AirVisual App. https://www.iqair.com/us/air-quality-app.

R. Iriondo. 2018. Machine Learning (ML) vs. AI and their Important Differences. *Towards AI* (revised 7 May 2020). https://medium.com/towards-artificial-intelligence/differences-between-ai-and-machine-learning-and-why-it-matters-1255b182fc6.

R. Iriondo. 2018. How Blockchain Will Transform Supply Chain Sustainability. *Towards AI* (revised 7 May 2020). https://medium.com/towards-artificial-intelligence/differences-between-ai-and-machine-learning-and-why-it-matters-1255b182fc6.

ITU. 2017. *ITU Expert group on household indicators (EGH). Background document 3. Proposal for a definition of Smartphone.* Geneva. https://www.itu.int/en/ITU-D/Statistics/Documents/events/egh2017/EGH%20 2017%20background%20document%203%20-%20Definition%20of%20smartphone.pdf.

———. 2019. *Turning Digital Technology Innovation into Climate Action.* Geneva. https://www.uncclearn.org/sites/ default/files/inventory/19-00405e-turning-digital-technology-innovation.pdf.

———. 2020. *ICTs 4 Disaster Management.* Geneva. https://www.itu.int/en/ITU-D/Emergency-Telecommunications/Pages/ICTs-4-DM.aspx.

IUCN. 2019. *Thailand introduces SMART tech to protect Asian elephants.* 3. September. https://www.iucn.org/news/ asia/201909/thailand-introduces-smart-tech-protect-asian-elephants.

J. Jambeck et al. 2015. Plastic waste inputs from land into the ocean. *Science.* Vol. 347 (6223). 13 February.

M.V. Japitana et al. 2019. Catchment Characterization to Support Water Monitoring and Management Decisions Using Remote Sensing. *Sustainable Environment Research (2019).* Vol. 29 (8). 11 April. https://doi.org/10.1186/s42834-019-0008-5.

F. Jiang et al. 2017. Artificial intelligence in healthcare: past, present and future. *Stroke and Vascular Neurology.* https://svn.bmj.com/content/2/4/230.full.

W. Jiang et al. 2015. Using social media to detect outdoor air pollution and monitor air quality index (AQI): a geo-targeted spatiotemporal analysis framework with Sina Weibo (Chinese Twitter). PloS one, 10(10), p.e0141185.

JioPhone 2. 2020. https://www.jio.com/en-in/jiophone2 (accessed 11 October 2020).

A. Johnson. 2020. How Artificial Intelligence is Aiding the Fight Against Coronavirus. *DataInnovation.* 13 March. https://www.datainnovation.org/2020/03/how-artificial-intelligence-is-aiding-the-fight-against-coronavirus/.

N. Joshi. 2019. How IoT and AI can enable Environmental Sustainability. *Forbes.* 4 September. https://www.forbes.com/sites/cognitiveworld/2019/09/04/how-iot-and-ai-can-enable-environmental-sustainability/#61e14b0268df.

———. 2019. Applications of immersive technologies in smart cities. *Forbes.* 5 August. https://www.forbes.com/ sites/cognitiveworld/2019/08/05/applications-of-immersive-technologies-in-smart-cities/.

B. Jovanovic and P. Rousseau. 2005. General Purpose Technologies. *The National Bureau of Economic Research.* https://www.nber.org/papers/w11093.

S. Keelery. 2020. Mobile phone internet user penetration India 2015–2023. *Statista*. 16 October. https://www.statista.com/statistics/309019/india-mobile-phone-internet-user-penetration/.

Kennedy Space Center. 2019. 60 Years Ago First Satellite Image of Earth. *The Payload Blog*. 7 August. https://www.kennedyspacecenter.com/blog/60-years-ago-first-satellite-image-of-earth#:~:text=On%20 August%207%2C%201959%2C%20the,scanning%20and%20photographing%20cloud%20cover.

N. Kshetri. 2016. *Big Data's Big Potential in Developing Economies: Impact on Agriculture, Health, and Environmental Security*. Wallingford, Oxon, UK: Centre for Agriculture and Biosciences International (CABI) Publishing.

———. 2021. Blockchain and Supply Chain Management. Elsevier. https://www.elsevier.com/books/blockchain-and-supply-chain-management/kshetri/978-0-323-89934-5.

I. Lacmanović. B. Radulović and D. Lacmanović. 2010. *Contactless payment systems based on RFID technology*. Conference Paper for the 33rd International Convention MIPRO. Croatia. 24–28 May.

N. Laskowski. 2016. *Delving into an enterprise IoT initiative? Read this first.* http://searchcio.techtarget.com/feature/ Delving-into-an-enterprise-IoT-initiative-Read-this-first.

R. Leblanc. 2019. E-Waste and the Importance of Electronics Recycling. The Balance Small Business. 25 June. https://www.thebalancesmb.com/e-waste-and-the-importance-of-electronics-recycling-2877783.

P. Liceras. 2019. Singapore experiments with its digital twin to improve city life. *Tomorrow City*. 20 May. https://www.smartcitylab.com/blog/digital-transformation/singapore-experiments-with-its-digital-twin-to-improve-city-life/.

Ledger Insights. 2018. China publishes blockchain service, identity regulations. https://www.ledgerinsights.com/ china-blockchain-regulations-identity-censorship/.

M.D. Le Sève et al. 2018. Delivering blockchain's potential for environmental sustainability. *ODI*. London. https://www.odi.org/sites/odi.org.uk/files/resource-documents/12439.pdf.

Y. Li et al. 2020. Big Data and Cloud Computing. Manual of Digital Earth. *Springer*, Singapore. https://link.springer. com/chapter/10.1007/978-981-32-9915-3_9.

I. Lochhead and N. Hedley. 2017. Mixed reality emergency management: bringing virtual evacuation simulations into real-world built environments. *International Journal of Digital Earth*. Vol. 12 (2).

J.D. Loftis. 2016. StormSense. *Virginia Institute of Marine Science*. https://www.vims.edu/people/loftis_jd/ StormSense/index.php.

H. Lovells. 2018. *Asia Data Protection and Cybersecurity Guide 2018*. https://f.datasrvr.com/fr1/818/81103/Hogan_ Lovells_Asia_Data_Protection_and_Cyber_Security_Guide_2018.pdf.

Low Emission Analysis Platform (LEAP). n.d. Marginal Abatement Cost Curve (MACC) Summary Reports. https://leap.sei.org/help/leap.htm#t=Views%2FMarginal_Abatement_Cost_Curve_(MACC)_Reports.htm.

X. Lu et al. 2016. Global Environmental Change. Unveiling hidden migration and mobility patterns in climate stressed regions: A longitudinal study of six million anonymous mobile phone users in Bangladesh. *ScienceDirect*. May. Vol. 38, pp. 1–7. https://www.sciencedirect.com/science/article/pii/ S0959378016300140.

D. D. Luxton. An Introduction to Artificial Intelligence in Behavioral and Mental Health Care, in: *Artificial Intelligence in Behavioral and Mental Health Care.* 2016. doi: 10.1016/B978-0-12-420248-1.00001-5. S. 25.

P. Mallya. 2014. Big Data: The invisible force reshaping our world. 15 March. http://www.thestar.com.my/Tech/ Tech-Opinion/2014/03/15/Big-Data-The-invisible-force-reshaping-our-world/.

T. Mann. 2019. Nokia VP: 5G Security Risks Are Huge. *sdxcentral.* 23 October. https://www.sdxcentral.com/ articles/news/nokia-vp-5g-security-risks-are-huge/2019/10/.

B. Marr. 2016. What Is The Difference Between Artificial Intelligence And Machine Learning? Forbes. 6 December. https://www.forbes.com/sites/bernardmarr/2016/12/06/what-is-the-difference-between-artificial-intelligence-and-machine-learning/#52916c542742.

Mathias. 2020. Robotic Process Automation – pragmatic solution or dangerous illusion?. BTOES Insights. 10 February. https://insights.btoes.com/risks-robotic-process-automation-pragmatic-solution-or-dangerous-illusion-1-1.

McKinsey Global Institute. 2018. *Smart Cities in Southeast Asia.* July. https://www.mckinsey.com/business-functions/operations/our-insights/smart-cities-in-southeast-asia.

C. Metz. 2017. Tech Giants Are Paying Huge Salaries for Scarce A.I. Talent. *The New York Times.* https://www.nytimes.com/2017/10/22/technology/artificial-intelligence-experts-salaries.html.

Microsoft. News Centre Europe. 2020. *How AI and satellite data are helping farmers waste less water.* https://news.microsoft.com/europe/features/how-ai-and-satellite-data-are-helping-farmers-waste-less-water/.

Microsoft. 2020. *Connected Devices: Microsoft's Heritage of Innovation.* https://www.microsoft.com/en-us/legal/ intellectualproperty/mtl/internet-of-things.aspx.

I. Mihajlovic. 2019. Everything You Ever Wanted to Know About Computer Vision. *Towards data science.* 25 April. https://towardsdatascience.com/everything-you-ever-wanted-to-know-about-computer-vision-heres-a-look-why-it-s-so-awesome-e8a58dfb641e.

S. Milrad. 2018. Satellite Imagery. Synoptic analysis and forecasting. *ScienceDirect.* https://www.sciencedirect.com/ topics/earth-and-planetary-sciences/satellite-imagery.

M. Minges et al. 2019. Disruptive technologies and their use in disaster risk management 2019. *ITUGET 2019.* Emergency Communication. Geneva. https://www.itu.int/en/ITU-D/Emergency-Telecommunications/ Documents/2019/GET_2019/Disruptive-Technologies.pdf.

C. Miskinis. 2018. What separates digital twin based simulations vs a reality that is augmented. *Challenge Advisory.* https://www.challenge.org/insights/digital-twin-vs-augmented-reality/.

Mongabay. 2015. Jokowi pledges Indonesia peatland 'revitalization' to stop the burning. 30 October. http://news.mongabay.com/2015/10/jokowi-pledges-greater-indonesia-peatland-revitalization-no-legal-breakthrough-yet/ (accessed 11 October 2020).

————. 2015. Singapore takes legal action against 5 Indonesian companies over haze. 1 October. http://news.mongabay.com/2015/10/singapore-takes-legal-action-against-5-indonesian-companies-over-haze/ (accessed 11 October 2020).

M. Moore. 2020. Average internet connection speed in selected Asia Pacific countries as of 1st quarter 2017. *Statista*. 22 July. https://statista.com/statistics/381388/asia-average-internet-connection-speed-by-country/#:~:text=Within%20the%20Asia%20Pacific%20region,4G%20download%20speed%20across%20Asia.

NASA Earth Observatory. 2020. *Airborne Nitrogen Dioxide Plummets Over China*. https://earthobservatory.nasa.gov/images/146362/airborne-nitrogen-dioxide-plummets-over-china (accessed 12 October 2020).

J. Nastu. 2020. Mercedes-Benz Pilots Blockchain Project to Track Emissions throughout Supply Chain. *Environment + Energy Leader*. https://www.environmentalleader.com/2020/01/mercedes-benz-pilots-blockchain-project-to-track-emissions-throughout-supply-chain/.

National Geographic. 2020. *GIS Geographical Information System*. https://www.nationalgeographic.org/encyclopedia/geographic-information-system-gis/12th-grade/.

National Ocean Service. 2020. *What is remote sensing?* https://oceanservice.noaa.gov/facts/remotesensing.html.

J. Nerad. 2020. Volvo Set to Challenge Tesla for Electric Car Supremacy. *Forbes*. https://www.forbes.com/sites/jacknerad2/2020/01/22/volvo-set-to-challenge-tesla-for-electric-car-supremacy/#613480e34011.

Neweurope. 2015. Indonesia's devastating forest fires are manmade. 9 November. https://www.neweurope.eu/article/indonesias-devastating-forest-fires-are-manmade/ (accessed 10 October 2020).

P. Newman. 2019. IoT Report: How Internet of Things technology growth is reaching mainstream companies and consumers. *Business Insider*. https://www.businessinsider.com/internet-of-things-report.

V.D. Ngo and N. Kshetri. 2017. *Business Opportunities and Barriers for Big Data in Vietnam*. Presentation for the 2017 PTC telecommunications conference. Hawaii. 15–18 January 2017. https://online.ptc.org/ptc17/program-and-attendees/proceeding.html?pid=242.

S. O'Dea. 2020. Smartphone users worldwide 2016–2021. *Statista*. 20. August. https://www.statista.com/statistics/330695/number-of-smartphone-users-worldwide/.

P. O'Donogue et al. 2016. Real-time anti-poaching tags could help prevent imminent species extinctions. *Journal of Applied Ecology*. Vol. 53 (1), pp. 5–10.

OECD. 2019. *PISA 2021. ICT Framework*. April. Paris. http://www.oecd.org/pisa/sitedocument/PISA-2021-ICT-framework.pdf.

———. 2020. *COVID-19 and the low-carbon transition: Impacts and possible policy responses*. 26 June. Paris. http://www.oecd.org/coronavirus/policy-responses/covid-19-and-the-low-carbon-transition-impacts-and-possible-policy-responses-749738fc/#section-d1e539.

———. 2020. *Digital security and privacy*. Paris. https://www.oecd.org/going-digital/topics/digital-security-and-privacy/.

———. 2020. *Glossary of statistical terms*. Paris. https://stats.oecd.org/glossary/.

G. Omale. 2018. Gartner Identifies Top 10 Strategic IoT Technologies and Trends. *Gartner*. https://www.gartner.com/en/newsroom/press-releases/2018-11-07-gartner-identifies-top-10-strategic-iot-technologies-and-trends.

E. Ortiz-Ospina. 2019. The rise of social media. *Our World in Data*. 18 September. https://ourworldindata.org/rise-of-social-media.

A. Ovanessoff and E. Plastino. 2017. How AI can drive South America's Growth. *Accenture*. https://www.accenture.com/_acnmedia/pdf-49/accenture-how-artificial-intelligence-can-drive-south-americas-growth.pdf.

A. Pamungkas. 2019. Indonesia: A smartphone app to manage household waste. *Deutsche Welle*. 6 December.

T.S. Perry. 2020. Satellites and AI Monitor Chinese Economy's Reaction to Coronavirus. *IEEE Spectrum*. 10 March. https://spectrum.ieee.org/view-from-the-valley/artificial-intelligence/machine-learning/satellites-and-ai-monitor-chinese-economys-reaction-to-coronavirus.

G. Petelin, M. Antoniou, and G. Papa. 2021. Multi-objective approaches to ground station scheduling for optimization of communication with satellites. Optimization and Engineering, pp. 1–38.

Pew Research Center. 2015. *Internet Access Strongly Related to Per Capita Income*. https://www.pewresearch.org/global/interactives/internet-usage/.

A. Peyton. 2019. China bosses blockchain and AI patents. *Fintech Futures*. 21 January. https://www.fintechfutures.com/2019/01/china-bosses-blockchain-and-ai-patents/.

PhD Essay. 2017. Socio-Economic and Environmental Impacts of Land Use Change. 7 February. https://phdessay.com/socio-economic-and-environmental-impacts-of-land-use-change/.

J.L. Ponz-Tienda et al. 2017. Marginal abatement Cost Curves (MACC): unsolved issues, and alternative proposals. *Green Energy and Technology*. June. https://www.researchgate.net/publication/316753935_Marginal_abatement_Cost_Curves_MACC_unsolved_issues_and_alternative_proposals.

Post-gazette.com. 2020. *A crowd in space: Tens of thousands of satellites planned for orbit*. 20 May. https://www.post-gazette.com/opinion/editorials/2020/05/20/space-satellites-crowded-junk-exploration-impact/stories/202002190057.

A. Prakash. 2018. Boiling Point. *Finance and Development*. September. Vol. 55 (3), p. 64.

PreventionWeb. 2019. *How can ICT help in disaster preparedness and response?* https://www.preventionweb.net/news/view/66649.

PwC. 2018. *Fourth Industrial Revolution for the Earth: Harnessing Artificial Intelligence for the Earth*. https://www.pwc.com/gx/en/news-room/docs/ai-for-the-earth.pdf.

———. 2019. Using AI to better manage the environment could reduce greenhouse gas emissions, boost global GDP by up to US $5 trillion and create up to 38m jobs by 2030. https://www.pwc.com/gx/en/news-room/press-releases/2019/ai-realise-gains-environment.html.

———. 2020. *How AI can enable a sustainable future. London*. https://www.pwc.co.uk/services/sustainability-climate-change/insights/how-ai-future-can-enable-sustainable-future.html.

———. 2020. *The Essential Eight*. https://www.pwc.com/gx/en/issues/technology/essential-eight-technologies.html.

Raptormaps.com. 2020. *Software and Aerial Inspection Services*. https://raptormaps.com/ (accessed 12 October 2020).

S. Ravi and P. Nagaraj. 2018. Harnessing the future of AI in India. *Brookings Institute*. 18 October. https://www.brookings.edu/research/harnessing-the-future-of-ai-in-india/.

RESET. 2020. *Drones: Propelling Sustainable Development*. https://reset.org/node/27732.

C. Riffle. 2017. What AI means for Sustainability. *GreenBiz*. 19 July. https://www.greenbiz.com/article/what-artificial-intelligence-means-sustainability.

J. Rotter and M. Price. 2019. The best emergency apps for wildfires, earthquakes, and other disasters. *CNET*. 1 November. https://www.cnet.com/news/the-best-emergency-apps-for-wildfires-hurricanes-earthquakes-and-other-disasters/.

M. Rouse. 2019. Definition-Internet of Things (IoT). techtarget. com, para. 1.

M. Sakurai and Y. Murayama. 2019. Information technologies and disaster management – Benefits and Issues. *ScienceDirect*. 2 July. https://www.sciencedirect.com/science/article/pii/S2590061719300122.

A. Salam. 2020. Internet of Things for Environmental Sustainability and Climate Change. Internet of Things for Sustainable Community Development. *Springer*, Cham. https://link.springer.com/chapter/10.1007/978-3-030-35291-2_2.

J. Sandino et al. 2018. Aerial Mapping of Forests Affected by Pathogens Using UAVs, Hyperspectral Sensors, and Artificial Intelligence. *Sensors 2018*. Vol. 18 (4). p. 944. https://www.clearbluesea.org/meet-fred/.

W. Sarni. 2020. *The Future of Water is Digital*. https://www.thesolutionsjournal.com/article/future-water-digital/.

E. Schmidt and J. Cohen. 2015. *Technology: Inventive artificial intelligence will make all of us better*. http://time.com/4154126/technology-essay-eric-schmidt-jared-cohen.

G. Schofield. 2013. Satellite tracking large numbers of individuals to infer population level dispersal and core areas for the protection of an endangered species. *Diversity and Distributions*. July. Vol. 19 (7).

S. Scoles. 2020. Satellite Data Reveals the Pandemic's Effects from Above. 9 April. https://www.wired.com/story/satellite-data-reveals-the-pandemics-effects-from-above/.

S. Sen. 2018. India moves to address AI talent supply gap, gets a leg-up from Google, Microsoft, Intel. *FactorDaily*. 18 January. https://factordaily.com/india-ai-talent-gap-google-microsoft/.

Sense Fly. 2020. *Using drones for aerial inspection for agriculture, with obvious applications for monitoring agricultural adaptation measures*. https://www.sensefly.com/.

R. Shuler. 2002. How does the Internet work? *Stanford University*. https://web.stanford.edu/class/msande91si/www-spr04/readings/week1/InternetWhitepaper.htm.

C.M. Shreve. 2014. Does mitigation save? Reviewing cost-benefit analyses of disaster risk reduction. *International Journal of Disaster Risk Reduction*. December. Vol. 10, Part A, pp. 213–235.

I. Singh. 2019. Lockheed Martin develops AI model for satellite imagery analysis. *Geoawesomeness*. 18 June. https://geoawesomeness.com/lockheed-martin-artificial-intelligence-model-satellite-imagery-analysis/.

M. Skourtos, A. Kontogianni and Ch. Tourkolias. 2013. Report on the Estimated Cost of Adaptation Options Under Climate Uncertainty. *The Climsave Project. Climate Change Integrated Assessment Methodology for Cross-Sectoral Adaptation and Vulnerability in Europe.* http://www.climsave.eu/climsave/doc/Report_on_adaptation_costs_under_uncertainty.pdf.

Statista. 2020. Smartphone penetration rate in China from 2015 to 2023. https://www.statista.com/statistics/321482/smartphone-user-penetration-in-china/ (accessed 25 November 2020).

P. Staunstrup. n.d. *Breakthrough for mobile telephony.* https://www.ericsson.com/en/about-us/history/products/mobile-telephony/breakthrough-for-mobile-telephony.

D. Stillman. 2014. What is a Satellite? *NASA* (updated 7 August 2017). https://www.nasa.gov/audience/forstudents/5-8/features/nasa-knows/what-is-a-satellite-58.html.

Stockholm University. 2020. *The nine planetary boundaries.* https://www.stockholmresilience.org/research/planetary-boundaries/planetary-boundaries/about-the-research/the-nine-planetary-boundaries.html.

P. Suprunov. 2018. How much does it cost to hire a blockchain developer? *Medium.* https://medium.com/practical-blockchain/how-much-does-it-cost-to-hire-a-blockchain-developer-16b4ffb372e5.

J. Sutton. 2016. Why Social Media and Sustainability Should Go Hand in Hand. *TriplePundit.* 8 June. https://www.triplepundit.com/story/2016/why-social-media-and-sustainability-should-go-hand-hand/25341.

Y. Sverdlik. 2020. Study: Data Centers Responsible for 1 Percent of All Electricity Consumed Worldwide. *DataCenterKnowldge.* 27 February. https://www.datacenterknowledge.com/energy/study-data-centers-responsible-1-percent-all-electricity-consumed-worldwide#:~:text=According%20to%20the%20study%2C%20in,electricity%20consumed%20that%20year%20worldwide.&text=The%20less%20energy%20a%20cloud,higher%20the%20provider's%20profit%20margin.

F.Z. Taibi and S. Konrad. 2018. Pocket Guide to NDCs under the UNFCCC. https://pubs.iied.org/sites/default/files/pdfs/migrate/G04320.pdf.

T-Mobile. 2019. *How The 5G Era Could Help Build A More Sustainable Future.* 21 October. https://www.forbes.com/sites/tmobile/2019/10/21/how-the-5g-era-could-help-build-a-more-sustainable-future/#6be2dcc4664f.

TechTerms. 2020. *ICT Definitions.* https://techterms.com/definition/ict.

———. 2020. *IT Definitions.* https://techterms.com/definition/it.

I. Todorović. 2020. EU preparing CO2 tax for products from other countries. *Balkan Green Energy News.* 29 July. https://balkangreenenergynews.com/eu-preparing-co2-tax-for-products-from-other-countries/.

F. Tenzer. 2020. Statistiken zu Smartphones. *Statista.* 8 September. https://de.statista.com/themen/581/smartphones/.

The Intergovernmental Panel on Climate Change. 2020. *IPCC Data.* https://www.ipcc.ch/data/.

The Intergovernmental Panel on Climate Change. 2014. *Climate Change 2014: Impacts, Adaptation, and Vulnerability. Part B: Regional Aspects.* Geneva.

The Ocean Cleanup. 2020. *High Quality Photos & Videos.* https://theoceancleanup.com/media-gallery/.

The Ocean Cleanup. 2020. *The largest Cleanup in History.* https://theoceancleanup.com/.

The Rodman Law Group. Distributed Ledger Technology. https://therodmanlawgroup.com/distributed-ledger-technology/ (accessed 18 April 2021)

L. L. Thomala. 2021. Penetration rate of social media in China 2013-2020. *Statista.* 19 February. https://www.statista.com/statistics/234991/penetration-rate-of-social-media-in-china/#:~:text=In%20 2020%2C%20the%20social%20media,approximately%20931%20million%20active%20users.

M. Trajtenberg. 2018. AI as the next General Purpose Technology: A Political-Economy Perspective. *National Bureau of Economic Research Working Paper Series.*

Tufts University. 2020. *Social Media Overview.* https://communications.tufts.edu/marketing-and-branding/social-media-overview/.

UK Office for National Statistics. (n.d.). *Measuring Sustainability Reporting Using Web Scraping and Natural Language Processing.*

UNFCCC. 2014. Lima Call for Action. https://unfccc.int/files/meetings/lima_dec_2014/application/pdf/auv_cop20_lima_call_for_climate_action.pdf.

UN ESCAP. 2015. *Utilizing space and GIS for effective disaster risk management - ESCAP's practices in Asia and the Pacific.* Committee on the Peaceful Uses of Outer Space: 2015, Fifty-eighth session. 10–19 June. Vienna, Austria. http://www.unoosa.org/pdf/pres/copuos2015/copuos2015tech21E.pdf.

———. 2018. *Key environment issues, trends, and challenges in the Asia-Pacific region. Environment and Development Series 2018.* Bangkok. https://www.unescap.org/sites/default/files/ST2849_Environment%20and%20 Development%20Series%202018_final.pdf.

———. 2019. *Summary of the Asia-Pacific Disaster Report 2019.* Bangkok. https://www.unescap.org/publications/asia-pacific-disaster-report-2019.

United Nations. 2014. *Big data and open data as sustainability tools. A working paper prepared by the Economic Commission for Latin America and the Caribbean.* Santiago. https://repositorio.cepal.org/bitstream/handle/11362/37158/1/S1420677_en.pdf.

———. 2020. *Big Data for sustainable development.* New York. https://www.un.org/en/sections/issues-depth/big-data-sustainable-development/.

Union of Concerned Scientists. 2019. *Climate Change and Agriculture. A Perfect Storm in Farm Country.* 20 March. https://www.ucsusa.org/resources/climate-change-and-agriculture.

UNOOSA. 2019. Ten years of the UN-SPIDER Beijing office. Vienna. https://www.unoosa.org/res/oosadoc/data/documents/2019/stspace/stspace_0_html/19-07423_UN_SPIDER_ebook_spreads.pdf S. van den Brink and J. Huisman. 2019. Approaches to responsible sourcing in mineral supply chains. *Resources, Conservation, and Recycling.* June. Vol. 145, pp. 389–398.

Verizon. 2019. *2019 Data Breach Investigations Report.* https://enterprise.verizon.com/resources/reports/dbir/.

Victoria State Government. 2020. *Teach with digital technologies.* https://www.education.vic.gov.au/school/teachers/teachingresources/digital/Pages/teach.aspx#:~:text=Digital%20technologies%20are%20electronic%20tools,of%20learning%20that%20uses%20technology.

J. Vincent. 2017. Tencent says there are only 300,000 AI engineers worldwide, but millions are needed. *The Verge.* https://www.theverge.com/2017/12/5/16737224/global-ai-talent-shortfall-tencent-report.

R. Vinuesa et al. 2020. The role of artificial intelligence in achieving the Sustainable Development Goals. *Nature Communications.* Vol. 11 (233). 13 January. https://www.nature.com/articles/s41467-019-14108-y.

Y. Waldeck. 2020. Smartphone penetration as share of population in South Korea 2015–2025. *Statista.* 17 July. https://www.statista.com/statistics/321408/smartphone-user-penetration-in-south-korea/.

F. Weiller. 2018. *Breaking The Psychological Barrier To Autonomous Vehicle Adoption.* 15 August. https://www.eeworldonline.com/breaking-the-psychological-barrier-to-autonomous-vehicle-adoption/.

A. Welfare. 2020. The Circular Economy and Sustainability powered by Blockchain. *Forbes.* 13 January. https://www.forbes.com/sites/forbestechcouncil/2020/01/13/the-circular-economy-and-sustainability-powered-by-blockchain/#7674db4eb8cf.

A. Wernick. 2018. Study: Climate change will bring more pests, crop losses. *Living on Earth.* 23 September. https://www.pri.org/stories/2018-09-23/study-climate-change-will-bring-more-pests-crop-losses#:~:text=A%20study%20published%20in%20the,increase%20with%20rising%20global%20temperatures.&text=A%20new%20study%20finds%20that,crop%20loss%20across%20the%20globe.

J. West and M. Bogers. 2013. Leveraging external sources of innovation: a review of research on open innovation. *Journal of Product Innovation Management.* November. Vol. 31 (4).

M. Wheatley. 2013. *Vietnam's Cities Use Big Data To Ward Off Traffic & Pollution. https://siliconangle.com/2013/08/16/vietnams-cities-use-big-data-to-ward-off-traffic-pollution/.*

WIPO. 2020. Climate-Friendly Information and Communication Technology. *The WIPO Green database.* Geneva. https://www3.wipo.int/wipogreen/en/news/2020/news_0021.html.

S. Wong. 2019. Smartphone penetration as share of population in China 2015–2023. *Statista.* 23 September. https://www.statista.com/statistics/321482/smartphone-user-penetration-in-china/.

Wong et al. 2018. *Mobile text alerts are an effective way of communicating emergency information to adolescents: Results from focus groups with 12- to 18-year-olds.* Contingencies and Crisis Management. *Wiley.* https://onlinelibrary.wiley.com/doi/pdf/10.1111/1468-5973.12185.

World Bank. 2010. *Energy Efficient Cities. Assessment Tools and Benchmarking Practices.* Washington, DC. http://documents1.World Bank.org/curated/en/602471468337215697/pdf/544330PUB0EPI01BOX034941 5B01PUBLIC1.pdf.

———. 2012. *The Role of Hydrometeorological Services in Disaster Risk Management.* Washington, DC. http://documents1.WorldBank.org/curated/en/960511468037565188/pdf/709420WP0P12910fHydrometerological0.pdf.

———. 2012. *Tools for Building Urban Resilience: Integrating Risk Information into Investment Decisions Pilot Cities Report – Jakarta and Can Tho.* Washington, DC. http://documents1.World Bank.org/curated/en/765581468234284004/pdf/714870WP0P124400JAKARTA0CAN0THO0WEB.pdf.

———. 2014. Third Flood Risk Management and Urban Resilience Workshop. 03 – 05 June. http://documents1.worldbank.org/curated/en/114171468245386234/pdf/906430WP0P13010roceedings0June02014.pdf.

———. 2015. *Enhancing the Weather Enterprise: A report on InterMET Asia 2015,* 23–25 April. Singapore. http://documents1.World Bank.org/curated/en/915781571828214136/pdf/Enhancing-the-Weather-Enterprise-A-Report-on-InterMET-Asia-2015-Singapore-23-to-25-April.pdf.

———. 2016. *Achieving Energy Savings by Intelligent Transportation Systems Investments in the Context of Smart Cities.* Washington, DC. http://documents1.World Bank.org/curated/en/913411469712967718/pdf/ACS18048-REVISED.pdf.

———. 2016. *Solving the Puzzle. Innovating to Reduce Risk.* Washington, DC. http://documents1.World Bank.org/curated/en/541521548734436712/pdf/Solving-the-Puzzle-Innovating-to-Reduce-Risk-Report.pdf.

———. 2016. *World Development Report 2016: Digital Dividends.* Washington, DC. World Bank. doi:10.1596/978-1-4648-0671-1. License: Creative Commons Attribution CC BY 3.0 IGO.

———. 2017. *Open access to risk information: Unlocking data, communicating risk.* Washington, DC. http://documents1.World Bank.org/curated/en/544661512542529342/pdf/Open-access-to-risk-information-unlocking-data-communicating-risk.pdf.

———. 2018. *After Shocks. Remodeling the Past for a resilient Future.* Washington, DC. http://documents1.World Bank.org/curated/en/808031525850704076/pdf/126048-REPLACEMENT-PUBLIC-Aftershocks.pdf.

———. 2018. *Blockchain and Emerging Digital Technologies for Enhancing Post-2020 Climate Markets.* Washington, DC. http://documents1.World Bank.org/curated/en/942981521464296927/pdf/124402-WP-Blockchainandemergingdigitaltechnologiesforenhancingpostclimatemarkets-PUBLIC.pdf.

———. 2018. *Using Satellite Imagery to Assess Impacts of Soil and Water Conservation Measures: Evidence from Ethiopia's Tana-Beles Watershed.* Washington, DC. http://documents1.World Bank.org/curated/en/210541517252830368/pdf/WPS8321.pdf.

———. 2018. *World Building Resilience through Innovation and Open Data in Sub Saharan Africa. Final Report.* 19 November. Washington, DC. http://documents1.World Bank.org/curated/en/785611543573934457/pdf/Final-Report-on-Building-Resilience-Through-Innovation-and-Open-Data-Program.pdf.

———. 2019. *Caribbean Digital Transformation Program (P171528). Project Information Document (PID).* http://documents1.worldbank.org/curated/en/202641563967196628/pdf/Concept-Project-Information-Document-PID-Caribbean-Digital-Transformation-Program-P171528.pdf.

———. 2019. *World Bank Open-Source Tools and Collaborative Platforms to Support National Greenhouse Gas Inventories in the AFOLU Sector. Workshop Report.* 28 March. Washington, DC. http://documents1.World Bank.org/curated/en/949571559665146166/pdf/Workshop-Report.pdf.

———. 2019. *World Bank Open-Source Tools and Collaborative Platforms to Support National Greenhouse Gas Inventories in the AFOLU Sector. Road Map Pilot in Three Countries and Verification Tool.* 28 March. Washington, DC. http://documents1.World Bank.org/curated/en/246411559665846894/pdf/Roadmap-Pilot-in-Three-Countries-and-Verification-Tool.pdf.

———. 2020. *Blockchain and Emerging DTs for Enhancing Post-2020 Climate Markets.* Washington, DC. https://olc.worldbank.org/system/files/Part%201-Blockchainandemergingdigitaltechnologiesforenhancingpostclimate markets-PUBLIC.pdf.

———. 2020. *Central African Backbone – APL1A.* Washington, D.C. https://projects.World Bank.org/en/projects-operations/project-detail/P108368.

———. 2020. *Digital Development: The World Bank provides knowledge and financing to help close the global digital divide, and make sure countries can take full advantage of the ongoing Digital Development revolution.* 27 October. Washington, DC. https://www.worldbank.org/en/topic/digitaldevelopment/overview#2.

———. 2020. Mobile cellular subscriptions (per 100 people) – Afghanistan. *Data.* https://data.worldbank.org/indicator/IT.CEL.SETS.P2?locations=AF.

———. 2020. Population, total. *Data.* https://data.worldbank.org/indicator/SP.POP.TOTL?end=2019&start=2015.

———. 2020. *Understanding Poverty. Digital Development.* Washington, DC. https://www.worldbank.org/en/topic/digitaldevelopment/overview#2.

———. 2020. *World Bank Outlook 2050. Strategic Directions Note. Supporting Countries to Meet Long-Term Goals of Decarbonization.* Washington, DC. https://openknowledge.worldbank.org/bitstream/handle/10986/33958/149871.pdf?sequence=3&isAllowed=y.

———. *Individuals using the Internet (% of population).* https://data.World Bank.org/indicator/IT.NET.USER.ZS (accessed 23 October 2020).

———. *Risk Assessments: Best Practices and Future Strategy.* Washington, DC. http://documents1.World Bank.org/curated/en/852261599811977901/pdf/Risk-Assessments-Best-Practices-and-Future-Strategy.pdf.

Worldometer. 2020. *World Population Projections.* https://www.worldometers.info/world-population/world-population-projections/.

World Population Review. 2020. *Internet Speeds by Country 2020.* https://worldpopulationreview.com/country-rankings/internet-speeds-by-country (accessed 15 October 2020).

Worldwildlife.org. 2018. *Coalition to End Wildlife Trafficking Online.* https://www.worldwildlife.org/pages/coalition-to-end-wildlife-trafficking-online.

Xinhua Net. 2018. Xinhua Headlines: Big data reshaping harvest for Chinese farmers. http://www.xinhuanet.com/english/2018-11/29/c_137640065_2.htm.

P. Yeung. 2019. How China's WeChat became a grim heart of illegal animal trading. *Wired.* 11 March.

Zee Media Bureau. 2020. Saudi Arabia has fastest 5G download speed, S Korea second --Full list of 15 countries. 20 October. https://zeenews.india.com/technology/saudi-arabia-has-fastest-5g-download-speed-s-korea-second-full-list-of-15-countries-2318863.html.

www.ingramcontent.com/pod-product-compliance
Lightning Source LLC
Chambersburg PA
CBHW050048220326
41599CB00045B/7323